Springer Theses

Recognizing Outstanding Ph.D. Research

Aims and Scope

The series "Springer Theses" brings together a selection of the very best Ph.D. theses from around the world and across the physical sciences. Nominated and endorsed by two recognized specialists, each published volume has been selected for its scientific excellence and the high impact of its contents for the pertinent field of research. For greater accessibility to non-specialists, the published versions include an extended introduction, as well as a foreword by the student's supervisor explaining the special relevance of the work for the field. As a whole, the series will provide a valuable resource both for newcomers to the research fields described, and for other scientists seeking detailed background information on special questions. Finally, it provides an accredited documentation of the valuable contributions made by today's younger generation of scientists.

Theses are accepted into the series by invited nomination only and must fulfill all of the following criteria

- They must be written in good English.
- The topic should fall within the confines of Chemistry, Physics, Earth Sciences, Engineering and related interdisciplinary fields such as Materials, Nanoscience, Chemical Engineering, Complex Systems and Biophysics.
- The work reported in the thesis must represent a significant scientific advance.
- If the thesis includes previously published material, permission to reproduce this must be gained from the respective copyright holder.
- They must have been examined and passed during the 12 months prior to nomination.
- Each thesis should include a foreword by the supervisor outlining the significance of its content.
- The theses should have a clearly defined structure including an introduction accessible to scientists not expert in that particular field.

Kelei Wang

Free Boundary Problems and Asymptotic Behavior of Singularly Perturbed Partial Differential Equations

Kelei Wang
Wuhan Inst. of Physics and Mathematics
Wuhan, Hubei
People's Republic of China

ISSN 2190-5053   ISSN 2190-5061 (electronic)
Springer Theses
ISBN 978-3-642-44248-3   ISBN 978-3-642-33696-6 (eBook)
DOI 10.1007/978-3-642-33696-6
Springer Heidelberg New York Dordrecht London

Mathematics Subject Classification: 35B25, 35B40, 35K57, 35R35, 49N60, 58E20, 92D25

Printed on acid-free paper

Springer is part of Springer Science+Business Media (www.springer.com)

Foreword

In recent decades, the phase separation phenomena arising from Bose–Einstein condensate theory and the competition model from population dynamics attracted a lot of attentions. It is observed that as the interactions between different species or condensates become large, their supports tend to be disjoint. In mathematics, these are modeled by a class of systems of partial differential equations (PDEs), with a parameter describing the interactions between different species. In the singular limit as this parameter goes to infinity (large interaction or strong competition, please see E.N. Dancer–Zhitao Zhang, Journal of Differential Equations, 2002), there appears a new class of free boundary problems. These free boundary problems play an important role in the understanding of the phase separation phenomena. These pose many difficult and interesting problems in mathematics.

Although these two models exhibit the same phase separation phenomena, we need to note a major difference between them: the PDE system arising in Bose–Einstein condensates has a variational structure (an energy associated to the PDE system), while the competing species system does not. Sometimes we need different techniques to treat these two problems. However, it has been conjectured for a long time that these two free boundary problems in the singular limit are the same one. This problem, despite its difficulty in mathematics, has many applications. For example, we know in general the dynamics of competing species is very complicated. However, if we know the above conjecture is true, then we can show that the dynamics of strongly competing species is very simple. In most cases, they will converge to a stationary configuration as time evolves.

This thesis is mainly devoted to the analysis of this free boundary problem and the asymptotic behavior of the two systems of singularly perturbed partial differential equations. Some new and deep techniques are developed to prove the uniqueness results for solutions of the singularly perturbed PDEs and the free boundary problems, and to establish the variational structure of the free boundary problems. Based on the understanding of this free boundary problems, it is proved rigorously that when the space dimension is one, strongly competing species converge to a stationary configuration as time goes to infinity. A method was also proposed to establish the connection between the two singular limits arising from Bose–Einstein condensation and competing species system, and this has been completed by subsequent research (joint with E.N. Dancer and Zhitao Zhang).

Beijing, China   Zhitao Zhang

Acknowledgements

First, I would like to express my deep gratitude to my thesis adviser, Professor Zhitao Zhang for introducing me to this field, and for his patience and guidance during my five years graduate studies.

I am also sincerely grateful to Professor E.N. Dancer for sharing his thoughts on problems related to this thesis. Furthermore, I also would like to thank my thesis committee members, Professor Daomin Cao, Professor Yanheng Ding, Professor Yuxia Guo, Professor Shujie Li and Professor Wenmin Zou, for kindly accepting to be the members of my thesis committee and their helpful comments on my thesis.

Also I would like to thank many friends, who made these five years so enjoyable. In addition, I would like to thank Yimin Sun for helps during preparing my thesis.

Last, but not the least, I would like to thank my parents and sister for their love and support.

Supplementary Note   Several parts of this thesis were published in the following articles:

1. K. Wang and Z. Zhang, Some new results in competing systems with many species. Ann. I. H. Poincaré-AN 2010; 27 (2):739–761. Copyright ©2009 Elsevier Masson SAS.
2. E.N. Dancer, K. Wang and Z. Zhang, Dynamics of strongly competing systems with many species, Trans. Amer. Math. Soc. 2012; 364, 961–1005. Copyright ©2011 the American Mathematical Society.
3. E.N. Dancer, K. Wang and Z. Zhang, The limit equation for the Gross-Pitaevskii equations and S. Terracini's conjecture, Journal of Functional Analysis 2012; 262 (2), 1087–1131. Copyright ©2011 Elsevier Masson SAS.

I have taken this opportunity to update some materials (most in the form of footnotes) and references in this thesis to reveal some subsequent progress about some problems discussed in this thesis after its completion.

Contents

Abbreviations

$\mathbb{R}^n$ The standard Euclidean space of dimension n

$B_R(x)$ The open ball with center x and radius R in $\mathbb{R}^n$

$Q_R(x, t)$ The parabolic cylinder $B_R(x) \times (t - R^2, t + R^2)$ in $\mathbb{R}^n \times (-\infty, +\infty)$

Ω A bounded domain in $\mathbb{R}^n$ with smooth boundary

$W^{1,p}(\Omega)$ The Sobolev space with the norm
$(1 \leq p \leq \infty)$

$$\|u\| := \left(\int_\Omega |u|^p + |\nabla u|^p \right)^{\frac{1}{p}}.$$

For $p = \infty$, this space is equivalent to the space of Lipschitz continuous functions on $\overline{\Omega}$

$H^1(\Omega)$ The Sobolev space $W^{1,2}(\Omega)$, which is a Hilbert space

$C^\alpha(\Omega)$ The space of Hölder continuous functions in Ω

dist_p The parabolic distance is defined as

$$\text{dist}_p\big((x_1, t_1), (x_2, t_2)\big) := |x_1 - x_2| + |t_1 - t_2|^{\frac{1}{2}}.$$

We can define the space of Hölder continuous functions with respect to the parabolic distance in a domain Q, which is still denoted by $C^\alpha(Q)$ in this thesis

$\mathcal{F}(u)$ The free boundary $\mathcal{F}(u) := \bigcup_i \partial\{u_i > 0\}$

$\text{reg}\,\mathcal{F}(u)$,   The regular set of free boundaries $\text{reg}\,\mathcal{F}(u)$ consists those points,

$\text{sing}\,\mathcal{F}(u)$ such that in a neighborhood of this point, the free boundary is a C^1 hypersurface. $\text{sing}\,\mathcal{F}(u)$ is the complement of the regular set

$U, (u_i)$ We denote vector valued functions by $U = (u_1, u_2, \ldots, u_M)$, sometimes also written as (u_i)

Chapter 1
Introduction

Abstract The Lotka–Volterra competition model from biological mathematics, which describes the dynamics of several species competing with each other, and the coupled Gross–Pitaevskii equations arising from Bose–Einstein condensates in theoretical physics, involve a class of systems of singularly perturbed elliptic or parabolic partial differential equations. The common feature in these problems is that there is a parameter κ. When κ is finite, we can observe partial overlap between different species (or condensates), and when κ goes to infinity (strong competition or large interaction), different species or condensates tend to be disjoint. This is known as the phase separation phenomena. This proposes a new class of free boundary problems, which is also related to the harmonic map into a singular space with non-positive curvature. In recent years, these problems have attracted a lot of interests. Many mathematicians, including L. Caffarelli, E.N. Dancer, F.H. Lin and S. Terracini have obtained a lot of deep results in this direction. In this chapter, we first introduce these problems. Then we recall some known results. In the last section, we list the main results in this thesis.

1.1 Competing Species System

The classical Lotka–Volterra competition model describes the behavior of several species competing with each other in a fixed domain. In general, this is modeled by a reaction-diffusion system:

$$\begin{cases} \dfrac{\partial u_i}{\partial t} - d_i \Delta u_i = f_i(u_i) - u_i \sum_{j \neq i} b_{ij} u_j, & \text{in } \Omega \times (0, +\infty), \\ u_i = \phi_i, & \text{on } \Omega \times \{0\}. \end{cases} \quad (1.1)$$

Here $b_{ij} > 0$, $d_i > 0$ are constants, and $1 \leq i, j \leq M$ with M the number of species (also the number of equations in this system). d_i represents the diffusion rate of the i-th species. Ω is a bounded domain in $\mathbb{R}^n$ ($n \geq 1$) with smooth boundary. u_i represents the density of the i-th species. The first term in the right-hand side of the equation, $f_i(u_i)$, describes the growth and self-constraints of the i-th species. It is usually taken to be ($a_i > 0$ is a constant) $f_i(u_i) = a_i u_i - u_i^2$. The competition between the

K. Wang, *Free Boundary Problems and Asymptotic Behavior of Singularly Perturbed Partial Differential Equations*, Springer Theses, DOI 10.1007/978-3-642-33696-6_1, © Springer-Verlag Berlin Heidelberg 2013

i-th species and the j-th species is modeled by the coupling term $b_{ij}u_iu_j$. When $t > 0$, we usually impose homogeneous Dirichlet boundary conditions or Neumann boundary conditions. (This requires compatible boundary conditions for the initial values ϕ_i.) We only consider positive solutions, that is, for all i, $u_i \geq 0$. (Positive solutions are always understood in this sense throughout this thesis.)

There is a lot of work on this reaction-diffusion system. However, most of these work are concerned with the case of two species (i.e., $M = 2$). One advantage in the case of two equations is, if we define a partial order relation for pairs of function, $(u_1, v_1) \geq (u_2, v_2)$ if $u_1 \geq v_1$ and $u_2 \leq v_2$, then a comparison principle holds. Hence the powerful method of monotone dynamical systems can be applied [40]. However, when the number of equations is greater than two, the situation is more complicated. For example, in [15, 16] E.N. Dancer and Yihong Du gave some interesting existence results when there are three species. In general, the above reaction-diffusion system can have very complicated dynamical behavior, especially when there are many species. For example, S. Smale [39] constructed some examples in the ODE case with chaotic dynamics. At present, there is still little understanding concerning the dynamics of (1.1), especially in the case with many species. The main difficulty is the lack of a variational structure of (1.1).

In recent years, there arises a lot of interests on the competing species system with strong competition. This is mainly concerned with the study of the following singularly perturbed parabolic system (or the corresponding elliptic system)

$$\frac{\partial u_i}{\partial t} - d_i \Delta u_i = a_i u_i - u_i^2 - \kappa u_i \sum_{j \neq i} b_{ij} u_j, \tag{1.2}$$

in particular, the properties of solutions when κ (a parameter describing the strength of competition) is very large, and the singular limit problem when κ goes to $+\infty$. Conti, Terracini and Verzini [10, 11], Caffarelli, Karakhanyan and Lin [2, 3], proved the regularity of solutions and the partial regularity of free boundaries in the singular limit κ, and the uniform Hölder continuity of solutions of (1.2) (for all $\kappa > 0$).

Conti, Terracini and Verzini discovered that in the singular limit as $\kappa \to +\infty$, different species tend to be spatially disjoint, and this limit satisfies a remarkable system of differential inequalities (in case of $d_i = 1$ for all i):

$$\begin{cases} \dfrac{\partial u_i}{\partial t} - \Delta u_i \leq a_i u_i - u_i^2 & \text{in } \Omega \times (0, +\infty); \\[3mm] \left(\dfrac{\partial}{\partial t} - \Delta\right)\left(u_i - \sum_{j \neq i} u_j\right) \geq a_i u_i - u_i^2 - \sum_{j \neq i}(a_j u_j - u_j^2) & \text{in } \Omega \times (0, +\infty); \\[3mm] u_i \geq 0, & \text{in } \Omega \times (0, +\infty); \\[3mm] u_i u_j = 0 & \text{in } \Omega \times (0, +\infty). \end{cases} \tag{1.3}$$

The above inequalities are understood in the H^1 sense, that is, for any $\phi(x,t) \geq 0$ which is smooth and has compact support, then

$$\iint_{\Omega \times (0,+\infty)} -\frac{\partial \phi}{\partial t} u_i + \nabla u_i \nabla \phi - (a_i u_i - u_i^2)\phi \, dx \, dt \leq 0.$$

Although Conti, Terracini and Verzini only considered elliptic problems, the treatment of parabolic problems is almost the same. In the following, for simplicity of presentations, we will consider a simplified system and the general case can be found in the above cited papers.

Consider

$$\begin{cases} \Delta u_i = \kappa u_i \sum_{j \neq i} b_{ij} u_j, & \text{in } \Omega, \\ u_i = \varphi_i, & \text{on } \partial \Omega. \end{cases} \tag{1.4}$$

Here $b_{ij} > 0$ are constants, and φ_i are nonnegative Lipschitz continuous functions defined on the boundary $\partial \Omega$. We assume that for $i \neq j$,

$$\varphi_i \varphi_j \equiv 0.$$

Compared to the problem (1.2), there is no zeroth order terms involving self-interactions in this problem. This does not affect the regularity issues of solutions. The free boundary problem in the singular limit as $\kappa \to +\infty$ of (1.4) is

$$\begin{cases} \Delta u_i \geq 0, & \text{in } \Omega, \\ \Delta\left(u_i - \sum_{j \neq i} \frac{b_{ij}}{b_{ji}} u_j\right) \leq 0, & \text{in } \Omega, \\ u_i = \varphi_i & \text{on } \partial \Omega, \\ u_i u_j = 0 & \text{in } \Omega. \end{cases} \tag{1.5}$$

Here, the free boundary is

$$\mathscr{F}(u) := \bigcup_i \partial\{u_i > 0\}.$$

The last condition in (1.5) implies that for different i, the supports of u_i are disjoint. Note that $\mathscr{F}(u)$ depends on the solution u and can not be prescribed, hence the name "free boundaries".

The result of Conti, Terracini and Verzini can be summarized as follows.

Theorem 1.1.1 [11] *For the problem* (1.4) *with the given boundary conditions specified as before, we have*

1. *For any fixed $\kappa > 0$, there exists a solution U_κ. For each i, $u_{i,\kappa} \in W^{1,\infty}(\Omega)$.*
2. *For any sequence $\kappa_l \to +\infty$, there exists a subsequence of U_{κ_l} converging strongly to a limit U in $(H^1(\Omega))^M$;*
3. *The limit U satisfies* (1.5).

Concerning the regularity of solutions, they proved two results. The first one is about the uniform continuity when $\kappa \to +\infty$.

Theorem 1.1.2 *For any $\alpha \in (0, 1)$ and i, $u_{i,\kappa}$ is uniformly bounded in $C^\alpha(\Omega)$ in κ.*

The second result is about the optimal regularity of the limit U.

Theorem 1.1.3 *Let $U \in (H^1(\Omega))^M$ satisfy (1.5) with the boundary condition as before, then for each i, $u_i \in W^{1,\infty}(\Omega)$.*

Caffarelli and Lin et al. [2] also proved these results by a different method. They also consider the corresponding problems in the parabolic case. The statement is exactly the same (of course, Hölder and Lipschitz continuity are measured with respect to the parabolic distance), and we do not repeat them here.

There is a remaining problem (see Sect. 2 in [4]), that is, in Theorem 1.1.2, can we improve the uniform Hölder continuity to be uniform Lipschitz continuity. Note that by Theorem 1.1.3, the limit U is Lipschitz continuous, which is optimal since the gradient of u_i has a jump when crossing the free boundary. For each κ, U_κ is also Lipschitz continuous. The problem is whether this Lipschitz continuity is uniform in κ. This is a very difficult problem. At present, there is only a positive answer of Conti, Terracini and Verzini in the case of two species [11], and no answer when there are many species in the system. We will give a partial answer to this problem in this thesis.

Conti, Terracini and Verzini also found that, if we assume the symmetric condition $b_{ij} = b_{ji}$, solutions to a variational problem also satisfy the system (1.5). More precisely, consider the metric space

$$\Sigma := \left\{ (u_1, u_2, \ldots, u_M) \in \mathbb{R}^M : u_i \geq 0, \text{ and if } i \neq j \; u_i u_j = 0 \right\}.$$

Under the intrinsic metric (i.e., the distance between two points in Σ is the minimal length of curves in Σ connecting them), this metric space has non-positive curvature (for more details about singular metric spaces with non-positive curvature, please see Gromov and Schoen [24]). For a map $U : \Omega \to \Sigma$, we can define an energy functional

$$D(u) := \int_\Omega \sum_i |\nabla u_i|^2. \tag{1.6}$$

By [24], this functional is convex with respect to the geodesic homotopy. Because any two points in a metric space with non-positive curvature can be connected by a unique geodesic, we can define the geodesic homotopy as follows.

Definition 1.1.4 The geodesic homotopy between two maps $u_0, u_1 : \Omega \mapsto \Sigma$ is a family of maps $u_t : \Omega \mapsto \Sigma, t \in [0, 1]$, such that $\forall x \in \Omega$, $u_t(x)$ is the unique point on the geodesic between $u_0(x)$ and $u_1(x)$, satisfying

$$dist\big(u_t(x), u_0(x)\big) = t \, dist\big(u_1(x), u_0(x)\big).$$

Here, dist is the intrinsic distance function of Σ.

Using the convexity of D, we can prove the uniqueness of its minimizer, cf. [10].

By choosing suitable comparison functions, Conti, Terracini and Verzini found that the minimizer of D satisfies (1.5). They then conjectured that if a map satisfies (1.5), it must be the minimizer of D. In [12], they proved this in a special case, that is, when there are there species and the space dimension is two. Their proof uses some special properties in two dimension, such as the isolatedness of singular points of free boundaries, and the fact that a solution of (1.5) can be patched together to give a harmonic function defined in Ω, which relies essentially on the topological property of the plane. They also used the connection between harmonic functions and complex analytic functions in two dimension. However, in higher dimensions, these results do not hold: the singular set of free boundaries could be a very complicated set of high dimension; it is impossible to patch solutions to get a harmonic function; there is no connection between harmonic functions and complex analytic functions any more. This conjecture is the second main problem studied in this thesis. We will give a complete proof for the general case.

To prove this conjecture, we need to understand the regularity of free boundaries in the singular limiting problem. Coming back to (1.5), this is a free boundary problem because in the interior of Ω, the boundary $\partial\{u_i > 0\}$ depends on the solution (u_i). Concerning the partial regularity of free boundaries in the symmetric case $b_{ij} = b_{ji}$, Conti, Terracini and Verzini [12] firstly obtained some partial results, and the full result was then established by Caffarelli, Karakhanyan and Lin in [2].

Theorem 1.1.5 [2] *If $U \in (H^1(\Omega))^M$ satisfies (1.5), where $b_{ij} = b_{ji}$, then the free boundary $\mathscr{F}(U)$ is a smooth hypersurface except a closed set of dimension $\leq n - 2$. This closed set is discrete if $n = 2$.*

Here the dimension is the Hausdorff dimension. For the parabolic system (1.3), the corresponding result is the following theorem.

Theorem 1.1.6 [2] *If $\forall U \in (C(\Omega \times (0, T)))^M$ satisfies (1.3), where $b_{ij} = b_{ji}$, then the free boundary $\mathscr{F}(U)$ is a smooth hypersurface except a closed set of parabolic Hausdorff dimension n. This closed set is discrete if $n = 1$.*

To prove these partial regularity results, they mainly use two techniques: the Clean Up lemma and an Almgren type monotonicity formula. These hold both in the elliptic case and the parabolic case. For simplicity, we only state them in the elliptic case and refer to [2] for the parabolic case.

Theorem 1.1.7 (Clean Up Lemma) *Given a $U \in (C(B_1(0)))^M$ satisfying (1.5), if the origin 0 lies on the free boundary $\mathscr{F}(U)$, and*

$$\lim_{r \to 0} \frac{1}{r} U(rx) = \left(\alpha x_1^+, \beta x_1^-, 0, \dots, 0\right),$$

where α and β are two positive constants, then in a small neighborhood of 0,

$$\sum_{i>2} u_i \equiv 0.$$

Theorem 1.1.8 (Almgren monotonicity formula) *Given a $U \in (C(B_1(0)))^M$ satisfying (1.5), where $b_{ij} = b_{ji}$, if the origin 0 lies on $\mathscr{F}(U)$, then*

$$N(r) := \frac{r \int_{B_r(0)} \sum_i |\nabla u_i|^2}{\int_{\partial B_r(0)} \sum_i u_i^2}$$

is nondecreasing in $r > 0$.

Sometimes the function $N(r)$ is called the frequency function, see [26].

With these regularity results in hand, we can prove that when $b_{ij} = b_{ji}$, the singular limiting system (1.3) and (1.5) do have a variational structure.

The above discussions show that, although the original system (1.2) does not have a variational structure, the singular limit as $\kappa \to +\infty$ does. Note that this fact was already noticed by Dancer and Du in [22] when the system has only two equations. As a consequence of this observation, we know (1.3) can not have a periodic (in time) solution. Then it is natural to ask (this is a conjecture of Dancer), whether the problem (1.2) has simple dynamics if κ is large. For example, can we prove that solutions to (1.2) will converge to a stationary solution as time goes to infinity? In biology, this means competing species will eventually tend to a stationary configuration. In [22], Dancer and Zhang answered this question affirmatively in the case of two species. But the case with many species (i.e., $M \geq 3$) has not been studied yet. Compared to the case with two species, there are many essential difficulties, which will be discussed in Sect. 1.3. This problem is the third main problem studied in this thesis.

1.2 Bose–Einstein Condensates

Another problem comes from the study of Bose–Einstein condensates. In 1997, Bose–Einstein condensation for a mixture of two different interacting atomic species with the same mass was firstly realized [27], exhibiting a partial overlap between the wave functions. Then a mathematical model was proposed to describe this phenomena [43]. In this model, two coupled Gross–Pitaevskii equations (a class of nonlinear Schrodinger equations) are used to analyze the interaction between two different condensates. The wave functions, ψ_1 and ψ_2, of two interacting condensates evolve according to

$$\begin{cases} i\hbar \dfrac{\partial \psi_1}{\partial t} = \left[-\dfrac{\hbar^2 \Delta}{2m_1} - \mu_1 + \lambda_1 |\psi_1|^2 \right] \psi_1 + k |\psi_2|^2 \psi_1, \\[2mm] i\hbar \dfrac{\partial \psi_2}{\partial t} = \left[-\dfrac{\hbar^2 \Delta}{2m_2} - \mu_2 + \lambda_2 |\psi_2|^2 \right] \psi_2 + k |\psi_1|^2 \psi_2, \end{cases} \tag{1.7}$$

where μ_j ($j = 1, 2$) represents the chemical potential of the j bosons. The interaction strengths, λ_j and k, are determined by the scattering lengths for binary collisions of like and unlike bosons. If $k > 0$, these two condensates are repulsive.

In this thesis, we are interested in the following coupled Gross–Pitaevskii equations, which arises if we want to find solutions to (1.7) with special ansatz such as standing wave solutions.

$$-\Delta u_i = f_i(u_i) - \kappa u_i \sum_{j \neq i} b_{ij} u_j^2. \tag{1.8}$$

Here $b_{ij} = b_{ji} > 0$ are constants. Under suitable boundary conditions, solutions to the problem (1.8) are critical points of the functional

$$J_\kappa(u) = \int_\Omega \frac{1}{2} \sum_i |\nabla u_i|^2 + \frac{\kappa}{4} \sum_{i \neq j} b_{ij} u_i^2 u_j^2 - \sum_i F_i(u_i). \tag{1.9}$$

In applications, we may take $F_i(u_i) = \frac{1}{2} a_i u_i^2 - \frac{1}{4} u_i^4$, but we can allow more general forms.

Similar to the competing species system, we can also consider the singular limit of (1.8) when $\kappa \to +\infty$. There are vast literatures on this problem. The following materials are mainly taken from the paper of Caffarelli and Lin, [3], and we refer to their paper for more references about this problem.

In [3], Caffarelli and Lin studied the convergence of minimizers of the functional (1.9) and the regularity problems of the limit. The main result is the following theorem.

Theorem 1.2.1 *Assume ϕ_i are Lipschitz continuous functions on $\partial \Omega$, such that $\phi_i \geq 0$ and $\phi_i \phi_j \equiv 0$ for $i \neq j$. For any $\kappa > 0$, the functional*

$$E_\kappa(u) = \int_\Omega \frac{1}{2} \sum_i |\nabla u_i|^2 + \frac{\kappa}{4} \sum_{i \neq j} b_{ij} u_i^2 u_j^2$$

has a minimizer in the class

$$\{U : u_i \in H^1(\Omega), \text{ and } u_i = \phi_i \text{ on } \partial \Omega\}.$$

Denote this minimizer as U_κ, then

1. *For any sequence of $\kappa_i \to +\infty$, there exists a subsequence of U_{κ_i} converging strongly to a limit U in $(H^1(\Omega))^M$;*
2. *U is the minimizer of the functional (1.6) in the class*

$$\{U : \Omega \mapsto \Sigma, u_i \in H^1(\Omega), u_i = \phi_i \text{ on } \partial\Omega\};$$

3. *U is Lipschitz continuous;*
4. *$\mathscr{F}(U)$ is a C^1 hypersurface, except a closed singular set of Hausdorff dimension $\leq n - 2$.*

Concerning the uniform Hölder continuity (with respect to κ) of solutions of (1.8), we have results similar to Theorem 1.1.2, see [36].

For positive solutions of (1.8), not necessarily minimizers, we can prove the following theorem.

Theorem 1.2.2 *Assume that there is a sequence of positive solutions U_{κ_i} to the problem (1.8) with the boundary conditions specified above. Then*

1. *For any sequence of $\kappa_i \to +\infty$, there exists a subsequence of U_{κ_i} converging strongly to a limit U in $(H^1(\Omega))^M$;*
2. *different components of U are disjoint, that is, $u_i u_j \equiv 0$ for $i \neq j$;*
3. *in the open set $\{u_i > 0\}$*

$$-\Delta u_i = f_i(u_i).$$

By [10], we know the minimizer U in Theorem 1.2.1 satisfies (1.5) (modulo some zeroth order terms). In [37], Terracini et al. conjectured that the limit U in Theorem 1.2.2 also satisfies (1.5) (modulo some zeroth order terms). Concerning this conjecture, this thesis will give a partial answer using the stationary condition.

For the parabolic case

$$\frac{\partial u_i}{\partial t} - \Delta u_i = a_i u_i - u_i^3 - \kappa u_i \sum_{j \neq i} b_{ij} u_j^2, \tag{1.10}$$

we can ask the same question: given a sequence of positive solutions U_κ, does their limit U satisfy (1.3). Our proof shows that this also has a positive answer.

1.3 Main Results

Most of this thesis is devoted to the study of the four main problems (and some related problems) introduced before.

In Chap. 2, we use the method of sub- and sup-solutions to prove the following result.

Theorem 1.3.1 *If $b_{ij} = b_{ji}$, then for any $\kappa \geq 0$, there exists a unique solution of (1.4).*

As was mentioned above, the existence part of this theorem has been established by Terracini et al. in [11]. They used the Leray–Schauder degree theory, but did not give the uniqueness of solutions. We also note that, although the application of the method of sub- and sup-solutions to nonlinear elliptic systems has a long history and is well known (for example, see [31]), problems similar to (1.4) has not been treated through this approach. In fact, the method of sub- and sup-solutions can not be used to give existence results for the general Lotka–Volterra competition system. However, in our case by utilizing the special structure of (1.4) and a simple observation, we can get the existence and uniqueness of solutions at the same time. Note that here the symmetric assumption $b_{ij} = b_{ji}$ is crucial for this method.

The same method can be applied to the parabolic case. Consider the parabolic counterpart of (1.4)

$$
\begin{cases}
\dfrac{\partial u_i}{\partial t} - \Delta u_i = -\kappa u_i \displaystyle\sum_{j \neq i} b_{ij} u_j, & \text{in } \Omega \times (0, +\infty), \\[2mm]
u_i = \varphi_i, & \text{on } \partial\Omega \times (0, +\infty), \\[1mm]
u_i = \phi_i, & \text{on } \Omega \times \{0\}.
\end{cases}
\tag{1.11}
$$

Here φ_i are given nonnegative Lipschitz continuous functions defined on $\partial\Omega$, and ϕ_i are given nonnegative Lipschitz continuous functions defined in Ω. They satisfy the compatible condition: $\phi_i = \varphi_i$ on $\partial\Omega$. By the method of sub- and sup-solutions, we will prove the following stability result for solutions of (1.11).

Theorem 1.3.2 *If $b_{ij} = b_{ji}$, then for any $\kappa \geq 0$, there exists a unique global solution of (1.11). As $t \to +\infty$, $U(t)$ converges to the solution of (1.4) in $C(\overline{\Omega})$.*

Next, we use the aforementioned observation and the Kato inequality to prove the uniform Lipschitz estimate introduced in Sect. 1.1. This is restricted to the symmetric case, i.e. $b_{ij} = b_{ji}$. This is still the optimal known result. Our method can be applied both to the elliptic case and the parabolic case.

Theorem 1.3.3 *If $b_{ij} = b_{ji}$, then there exists a constant $C > 0$ independent of κ, such that for solutions $(u_{i,\kappa})$ of (1.4), we have*

$$\sup_{\Omega} |\nabla u_{i,\kappa}| \leq C.$$

Theorem 1.3.4 *If $b_{ij} = b_{ji}$, then there exists a constant $C > 0$ independent of κ, such that for solutions $(u_{i,\kappa})$ of (1.11), we have*

$$\sup_{\Omega \times [0, +\infty)} Lip(u_{i,\kappa}) \leq C.$$

Here the Lipschitz constant is measured with respect to the parabolic distance. In these two results, we need more conditions on the (initial-)boundary values. This will be specified in Sect. 2.3. There we will also give some interior estimates.

In Chap. 3, we study some properties of the singular limit problem. For solutions of (1.5) (in the case of $b_{ij} = b_{ji}$), by the partial regularity result in [2], we know the Hausdorff dimension of free boundaries is $n - 1$. However, this does not say anything about the $n - 1$ dimensional Hausdorff measure of free boundaries. In Sect. 3.1, by developing the compactness method of Qing Han and Fanghua Lin [26], we give an interior estimate of $n - 1$ dimensional Hausdorff measure of free boundaries. For any $N > 0$, define

$$H_N^1 := \left\{ u : B_1(0) \to \Sigma, \text{ satisfying (1.5) except the boundary conditions therein,} \right.$$

$$\left. \text{and } \frac{\int_{B_1(0)} \sum_i |\nabla u_i^2|}{\int_{\partial B_1(0)} \sum_i u_i^2} \le N, \int_{\partial B_{\frac{1}{2}}(0)} \sum_i u_i^2 = 1 \right\}.$$

Then we have

Theorem 1.3.5 *There exists a constant $C(N, n)$ depending only on N and the dimension n, for any $U \in H_N^1$,*

$$H^{n-1}\left(B_{\frac{1}{2}}(0) \cap \mathscr{F}(u)\right) \le C(N, n).$$

Here H^{n-1} denotes the $n - 1$ dimensional Hausdorff measure.

This estimate also holds for local energy minimizing map into Σ. This is because they enjoy the same properties such as Almgren type monotonicity formula. This method can also be used to get a uniform estimate of the $n - 1$ dimensional Hausdorff measure of level sets.

In Sect. 3.2, we study the second main problem introduced in Sect. 1.1, that is, the variational structure of (1.5) (when $b_{ij} = b_{ji}$). We prove the following theorem.

Theorem 1.3.6 *Given boundary conditions φ as in (1.5), there exists a unique solution of (1.5). Moreover, this solution is the minimizer of the energy functional (1.6).*

The idea is to compare the energy of a solution of (1.5) and the energy minimizer v (with the same boundary conditions). This needs us to construct the geodesic homotopy between them and calculate the derivative of the energy functional at $t = 0$. By formal calculations using integration by parts, we can transfer those terms in this derivative into integrals on free boundaries. These integrals can be canceled by using the balancing conditions on free boundaries, which are implied by (1.5). However, since free boundaries $\partial\{u_i > 0\}$ may have singularities, we do not know whether we can integrate by parts in the domain $\{u_i > 0\}$. To make the above arguments rigorous, we truncate u_i and v_i, and use smooth levels to replace the possible singular free boundaries. After obtaining some uniform control on integrals on these level sets, we can pass to the limit to get integrals on free boundaries.

We also proved the uniqueness of solutions of initial-boundary value problems for the singular parabolic system (1.3). Due to the singular nature of this parabolic system, this uniqueness result is rather surprising. This result can also be considered as an L^1 contraction principle. This result, combined with some results about (1.10), can be used to derive the variational structure of (1.3). That is, solutions of (1.3) satisfies an energy inequality.

Theorem 1.3.7 *There exists a unique solution for the problem* (1.3). *Moreover, as* $t \to +\infty$, *this solution converges to the unique solution of Eq.* (1.5).

Next, we consider the third problem posed before. In Chap. 4, we study the one dimensional case of (1.3) in great details. This is the following system

$$
\begin{cases}
\dfrac{\partial u_i}{\partial t} - \dfrac{\partial^2 u_i}{\partial x^2} \leq a_i u_i - u_i^2 & \text{in } (0,1) \times (0,+\infty); \\[2mm]
\left(\dfrac{\partial}{\partial t} - \dfrac{\partial^2}{\partial x^2} \right) \left(u_i - \sum_{j \neq i} u_j \right) \geq a_i u_i - u_i^2 - \sum_{j \neq i} \left(a_j u_j - u_j^2 \right) \\[2mm]
\qquad \text{on } (0,1) \times (0,+\infty); \\[2mm]
u_i = \phi_i, \quad \text{on } [0,1] \times \{0\}; \\[2mm]
u_i u_j = 0 \quad \text{on } [0,1] \times (0,+\infty).
\end{cases}
\tag{1.12}
$$

Here $i = 1, 2, \ldots, M$, ϕ_i are given Lipschitz continuous functions on $[0,1]$, satisfying

$$\phi_i \geq 0, \quad \text{and if} \quad i \neq j \quad \phi_i \phi_j \equiv 0.$$

For simplicity, we take $b_{ij} \equiv 1$ in the original problem (1.1). The general case can be treated by the same method. The main result of Chap. 4 can be summarized in the following theorem.

Theorem 1.3.8 *Let*

$$m(T) := \textit{The number of connected components of } \bigcup_i \{u_i > 0\} \cap \{t = T\},$$

then

1. $m(T)$ *is non-increasing in* T;
2. *there exist at most countable* $T_1 > T_2 > \cdots$, *satisfying* $\lim_{k \to \infty} T_k = 0$, *such that these points are exactly those times where the value of* $m(T)$ *jumps*;
3. *for* $t \in (T_{i+1}, T_i)$ *(here we take the convention that* $T_0 = +\infty$*), after re-indexing* i *and ignoring those* u_i *which are identically* 0 *in this time interval, we can define*

$$u := u_1 - u_2 + u_3 - u_4 + \cdots.$$

u satisfies

$$\left(\frac{\partial}{\partial t} - \frac{\partial^2}{\partial x^2} \right) u = f. \tag{1.13}$$

Here $f := a_1 u_1 - u_1^2 - (a_2 u_2 - u_2^2) + \cdots$ is a Lipschitz continuous function.

4. *In $[0, 1] \times (T_1, +\infty)$, we have $\lim_{t \to +\infty} u_i(x, t) = v_i(x)$, where $v := (v_1, v_2, \ldots)$ is a stationary solution of (1.12). Moreover, either $v \equiv 0$ or v has exactly $m(T_1)$ nonzero components.*

This theorem shows that, even though the initial values may have infinitely many connected components of $\{u_i > 0\}$, after an arbitrarily small time, there will be only finitely many connected components (consequently, finitely many species) left. This result also shows that, after some time, no species will extinct, and as $t \to +\infty$, the solution converges to a stationary solution. Another interesting phenomena is that no species can extinct at infinite time. In other words, if a species is very small (in population) compared with other species, it will extinct in finite time.

With these understandings of the singular limit problem, we continue to study the dynamics of the strongly competing system with many species:

$$\begin{cases} \dfrac{\partial u_i}{\partial t} - \dfrac{\partial^2 u_i}{\partial x^2} = a_i u_i - u_i^2 - \kappa u_i \displaystyle\sum_{j \neq i} b_{ij} u_j & \text{in } [0, 1] \times (0, +\infty), \\[4mm] u_i = \phi_i & \text{on } [0, 1] \times \{0\}. \end{cases} \tag{1.14}$$

The results in Theorem 1.3.8 say that, the singular limit of (1.14) as $\kappa \to +\infty$ has a variational structure, and the solution converges to a stationary solution as time goes to infinity. We prove that, under some natural non-degeneracy assumptions (see Sect. 6.1 for more details), the dynamics of (1.14) is very simple.

Theorem 1.3.9 *Under suitable assumptions, if κ is sufficiently large, any solution of Eq. (1.14) converges to a stationary solution as t goes to infinity.*

Dancer and Zhang [22] have proved the corresponding result in the case of two species. To generalize their ideas to the case with many species, the main difficulty lies in the following fact: in the case of two species, we can subtract two equations in the system to get a single equation, but this is impossible in the case with many species. This is due to the possibility that many (more than 3) species could compete each other at the same place. However, we find that when the space dimension is 1, due to some topological restrictions, in most places there are at most two species.

In fact, in the place where there are only two species non-vanishing in the singular limit of $\kappa \to +\infty$, when κ is sufficiently large, compared with these two dominated species, the population of other species is very small. In fact, the impact of these species can be neglected. We call this result "Approximate Clean Up" lemma. This lemma is based on the "Clean Up" lemma of Caffarelli, Karakhanyan and Lin (see [2]). This lemma holds in any dimension $n \geq 1$. We also give some variant of this lemma such as the linearized version and boundary versions, etc. Since these results are rather technical, here we do not state them and we refer the reader to Chap. 5 for more details. The main results in Chap. 5 include Theorem 5.1.1 (the "Clean Up" lemma in the singular limit), Theorem 5.2.2 (the approximate "Clean Up" lemma), Proposition 5.3.3 (the linearized version), Proposition 5.4.1 (the case near fixed boundaries) and Proposition 5.4.2 (the linearized version near fixed boundaries).

In [22], in order to get a uniform bound of solutions of the linearized equations of (1.2) from their L^2 bound, Dancer and Zhang used the Kato inequality. It seems that this method can not work in the case with many species, except when we have the symmetric assumption $b_{ij} = b_{ji}$. The "Approximate Clean Up" lemmas we developed here can be used to overcome this difficulty. When the space dimension is 1, there may exist other methods to prove these results directly. But our methods can be applied to higher dimensional problems and we believe that they are more important.

We mainly consider the non-existence of periodic solutions of strongly competing species system when the space dimension is 1. But our method can be easily generalized to high dimensions, if we assume some non-degeneracy conditions. However, although most of these conditions hold unconditionally in 1 dimension, they are not so natural in high dimensions. We hope that, if we do not assume these conditions, we can still prove a weaker result: when κ is sufficiently large, any periodic solution of Eq. (1.2) must stay in a small neighborhood of a stationary solution of the singular limit. This means, this periodic solution varies slowly as time goes on. So it looks almost like a stationary solution.

In the last chapter, we study the problem posed in Sect. 1.2. S. Terracini et al. conjectured that the limit equation of (1.8) should be (here we assume the domain is the unit ball)

$$
\begin{cases}
-\Delta u_i \le f_i(u_i), & \text{in } B_1(0), \\
-\Delta\left(u_i - \sum_{j \ne i} u_j\right) \ge f_i(u_i) - \sum_{j \ne i} f_j(u_j), & \text{in } B_1(0), \\
u_i \ge 0 & \text{in } B_1(0), \\
u_i u_j = 0 & \text{in } B_1(0), \text{ if } i \ne j.
\end{cases}
\tag{1.15}
$$

In Chap. 7, we first prove the following result.

Theorem 1.3.10 *Let u_κ be a sequence of uniformly bounded solutions of Eq. (1.8) in $B_1(0)$, that is, there exists a constant $C > 0$ independent of κ, such that $\max_i \sup_{B_1(0)} u_{i,\kappa} \le C$. Then as $\kappa \to +\infty$, the limit of u_κ, u satisfies (1.15).*

We then study the corresponding parabolic problem.

$$
\frac{\partial u_i}{\partial t} - \Delta u_i = f_i(u_i) - \kappa u_i \sum_{j \ne i} b_{ij} u_j^2, \quad \text{in } B_1(0) \times (-1, 1).
\tag{1.16}
$$

Inspired by the elliptic case, it is natural to conjecture that the limit problem as $\kappa \to +\infty$ should be

$$
\begin{cases}
\dfrac{\partial u_i}{\partial t} - \Delta u_i \le f_i(u_i), & \text{in } Q_1(0), \\[2mm]
\left(\dfrac{\partial u_i}{\partial t} - \Delta\right)\left(u_i - \sum_{j \ne i} u_j\right) \ge f_i(u_i) - \sum_{j \ne i} f_j(u_j), & \text{in } Q_1(0), \\[2mm]
u_i \ge 0 & \text{in } Q_1(0), \\[2mm]
u_i u_j = 0 & \text{in } Q_1(0).
\end{cases}
\tag{1.17}
$$

If we add a homogeneous Dirichlet boundary conditions on the problem (1.16), then this equation defines the gradient flow of the functional (1.9). From this viewpoint, we can get an energy inequality for solutions of (1.16). In [5], it was proved that this energy inequality can be passed to the limit as $\kappa \to +\infty$. So, without loss of generality, we can assume the solution of (1.16), u_κ satisfies

$$
\iint_{Q_1(0)} |\nabla u_\kappa|^2 + |\frac{\partial u_\kappa}{\partial t}|^2 \le C,
\tag{1.18}
$$

and for any $t \in (-1, 1)$

$$
\int_{B_1 \times \{t\}} |\nabla u_\kappa|^2 + \kappa \sum_{i \ne j} u_{i,\kappa}^2 u_{j,\kappa}^2 \le C.
\tag{1.19}
$$

Here the constant $C > 0$ is independent of κ. Under these assumptions, we can prove

Theorem 1.3.11 *Let u_κ be a sequence of uniformly bounded solutions of Eq.* (1.16) *in $Q_1(0)$, that is, there exists a constant $C > 0$ independent of κ, such that $\max_i \sup_{Q_1(0)} u_{i,\kappa} \le C$. Assume u_κ satisfies* (1.18) *and* (1.19). *Then as $\kappa \to +\infty$, u_κ converges to a solution u of Eq.* (1.17).

The proof of these two theorems are similar. A crucial step is to prove that, in a small neighborhood of a regular point on the free boundary, the two component of U, u_1 and u_2, can be continued across the free boundary so that $u_1 - u_2$ become a solution of a single (elliptic or parabolic) equation. This needs us to prove that the normal derivatives of these two components along the free boundary coincide. In order to prove this, we use a new condition satisfied by critical points of a functional. Usually we take variations in the target space to derive the Euler–Lagrange equations satisfied by the critical point of a functional. However, we can also use variations in the domain of definition, through the one parameter transformation group generated by a smooth vector field with compact support. Using this we can derive an identity for critical points, which we call a stationary condition (this can be seen as a generalization of the first integral for the ordinary differential equation problem in dimension 1). In many other problems such as harmonic maps, this condition can be used to derive some important monotonicity formulas.

However, in the current singular limit problem, due to the appearance of free boundaries, we need more efforts to make this derivation of the stationary condition rigorous. In this thesis, we take the following viewpoint on this problem. Note that solutions to the limit problem, U, are limits of U_κ, which are solutions to the original problem with finite κ. U_κ can be proved to be smooth, and satisfies the stationary condition by using the above argument. Then we can let $\kappa \to +\infty$ in this stationary condition and get the stationary condition for the singular limit. Finally, by some integration by parts, we can transfer the stationary condition into some integrals on free boundaries. This then implies some balance conditions on the free boundary and finishes the proof of Theorem 1.3.10.

For parabolic problems, the basic idea is the same, that is, to derive some monotonicity formula by taking limit in κ and using integration by parts to get the balance conditions on the free boundary. Hugo Tavares and Susanna Terracini in [42] essentially proved the above Theorem 1.3.10.[1] They also used the monotonicity formula, but not the stationary condition mentioned above.

---

[1] This paper has been published after the completion of this thesis. For readers' convenience, we have updated the references. Note that the argument in Chap. 7 is not complete. For example, there we assumes the smoothness of free boundaries, thus we do not touch the partial regularity problem of free boundaries. These problems have been solved in subsequent researches. We refer the reader to [19, 42] for further details, in particular, for the problem on the gap phenomena of density functions and the non-existence of multiplicity one point on the free boundary.

Chapter 2
Uniqueness, Stability and Uniform Lipschitz Estimates

Abstract In this chapter, we first prove the uniqueness of solutions to the Dirichlet boundary value problem (1.4) by the sub- and super-solution method. In Sect. 2.2, we use the same method to prove the stability of solutions to the corresponding parabolic initial-boundary value problem. Finally, by the same idea, we prove the uniform Lipschitz estimates for solutions to these two problems, under suitable boundary conditions.

2.1 A Uniqueness Result for the Elliptic System

In this section, we prove the uniqueness of solutions to the following Dirichlet boundary value problem.

$$\begin{cases} \Delta u_i = \kappa u_i \sum_{j \neq i} b_{ij} u_j, & \text{in } \Omega, \\ u_i = \varphi_i, & \text{on } \partial\Omega. \end{cases} \tag{2.1}$$

Here $b_{ij} \geq 0$ are constants, satisfying $b_{ij} = b_{ji}$. φ_i are given nonnegative Lipschitz continuous functions on $\partial\Omega$. We prove the following theorem.

Theorem 2.1.1 *For any $\kappa \geq 0$, there exists a unique solution to the problem* (2.1).

We use the following iteration scheme to prove the uniqueness of solutions for (2.1). First, we know the following harmonic extension is possible:

$$\begin{cases} \Delta u_{i,0} = 0, & \text{in } \Omega, \\ u_{i,0} = \varphi_i, & \text{on } \partial\Omega, \end{cases} \tag{2.2}$$

that is, this equation has a unique positive solution $u_{i,0} \in C^2(\Omega) \cap C^0(\bar{\Omega})$ by Theorem 4.3 of [25].

Then the iteration can be defined as:

$$\begin{cases} \Delta u_{i,m+1} = \kappa u_{i,m+1} \sum_{j \neq i} u_{j,m}, & \text{in } \Omega, \\ u_{i,m+1} = \varphi_i & \text{on } \partial\Omega. \end{cases} \tag{2.3}$$

K. Wang, *Free Boundary Problems and Asymptotic Behavior of Singularly Perturbed Partial Differential Equations*, Springer Theses, DOI 10.1007/978-3-642-33696-6_2, © Springer-Verlag Berlin Heidelberg 2013

This is a linear equation. It satisfies the maximum principle, so the existence and uniqueness of the solution $u_{i,m+1} \in C^2(\Omega) \cap C^0(\bar{\Omega})$ is clear (cf. Theorem 6.13 in [25]).

Now concerning these $u_{i,m}$, we have the following result.

Proposition 2.1.2 *In Ω*

$$u_{i,0}(x) > u_{i,2}(x) > \cdots > u_{i,2m}(x) > \cdots > u_{i,2m+1}(x) > \cdots > u_{i,3}(x) > u_{i,1}(x).$$

Proof We divide the proof into several claims.

Claim 1 $\forall i, m, u_{i,m} > 0$ in Ω.

Because $\sum_{j \neq i} u_{j,0} > 0$ in Ω, the equation (2.3) satisfies the maximum principle. Because the boundary value $\varphi_i \geq 0$, $u_{i,1} > 0$ in Ω. By induction, we see the claim holds true for all $u_{i,m}$.

Claim 2 $u_{i,1} < u_{i,0}$ in Ω.

From the equation, now we have

$$\begin{cases} \Delta u_{i,1} \geq 0, & \text{in } \Omega, \\ u_{i,1} = u_{i,0}, & \text{on } \partial\Omega, \end{cases}$$

so we get $u_{i,1} < u_{i,0}$ by the comparison principle.

In the following, we assume the conclusion of the proposition is valid until $2m + 1$, that is in Ω

$$u_{i,0} > \cdots > u_{i,2m} > u_{i,2m+1} > u_{i,2m-1} > \cdots > u_{i,1}.$$

Then we have the following.

Claim 3 $u_{i,2m+1} \leq u_{i,2m+2}.$

By (2.3), we have

$$\Delta u_{i,2m+2} \leq \kappa u_{i,2m+2} \sum_{j \neq i} u_{j,2m}. \tag{2.4}$$

$$\Delta u_{i,2m+1} = \kappa u_{i,2m+1} \sum_{j \neq i} u_{j,2m}. \tag{2.5}$$

Because $u_{i,2m+1}$ and $u_{i,2m+2}$ have the same boundary value, comparing (2.4) and (2.5), by the comparison principle again we obtain that $u_{i,2m+1} \leq u_{i,2m+2}.$

Claim 4 $u_{i,2m+2} \leq u_{i,2m}.$

This can be seen by comparing the equations they satisfy:

$$\begin{cases} \Delta u_{i,2m+2} = \kappa u_{i,2m+2} \sum_{j \neq i} u_{j,2m+1}, \\ \Delta u_{i,2m} = \kappa u_{i,2m} \sum_{j \neq i} u_{j,2m-1}. \end{cases}$$

By assumption, we have $u_{j,2m+1} \geq u_{j,2m-1}$, so the claim follows from the comparison principle again.

Claim 5 $u_{i,2m+3} \geq u_{i,2m+1}$.

This can be seen by comparing the equations they satisfy:

$$\begin{cases} \Delta u_{i,2m+3} = \kappa u_{i,2m+2} \sum_{j \neq i} u_{j,2m+2}, \\ \Delta u_{i,2m+1} = \kappa u_{i,2m+1} \sum_{j \neq i} u_{j,2m}. \end{cases}$$

By Claim 4, we have $u_{j,2m} \geq u_{j,2m+2}$, so the claim follows from the comparison principle again.

Now we know that there exist two family of functions u_i and v_i, such that $\lim_{m \to \infty} u_{j,2m}(x) = u_j(x)$ and $\lim_{m \to \infty} u_{j,2m+1}(x) = v_j(x)$, $\forall x \in \Omega$. Moreover, by standard elliptic estimates, we know this convergence is smooth in Ω and uniformly on $\overline{\Omega}$. So by taking the limit in (2.3), we obtain the following equations:

$$\begin{cases} \Delta u_i = \kappa u_i \sum_{j \neq i} v_j, \\ \Delta v_i = \kappa v_i \sum_{j \neq i} u_j. \end{cases} \tag{2.6}$$

Because $u_{i,2m+1} \leq u_{j,2m}$, by taking limit, we also have

$$v_i \leq u_i. \tag{2.7}$$

Now summing (2.6), we have

$$\begin{cases} \Delta\left(\sum_i u_i\right) = \kappa \sum_i \left(u_i \sum_{j \neq i} v_j\right), \\ \Delta\left(\sum_i v_i\right) = \kappa \sum_i \left(v_i \sum_{j \neq i} u_j\right). \end{cases} \tag{2.8}$$

It is easily seen that

$$\sum_i \left(u_i \sum_{j \neq i} v_j\right) = \sum_i v_i \left(\sum_{j \neq i} u_j\right),$$

so we must have $\sum_i u_i \equiv \sum_i v_i$ because they have the same boundary value. This means, by (2.7), $u_i \equiv v_i \in C^2(\Omega) \cap C^0(\bar{\Omega})$. In particular, they satisfy (2.1). This proves the existence part of Theorem 2.1.1.   □

Proposition 2.1.3 *If there exist another positive solution w_i of (2.1), we must have $u_i \equiv w_i$.*

Proof We will prove $u_{i,2m} \geq w_i \geq u_{i,2m+1}, \forall m$, and then the proposition follows immediately. We divide the proof into several claims.

Claim 1 $w_i \leq u_{i,0}$.
   This is because

$$\begin{cases} \Delta w_i \geq 0, & \text{in } \Omega, \\ w_i = u_{i,0}, & \text{on } \partial\Omega. \end{cases}$$

Claim 2 $w_i \geq u_{i,1}$.
   This is because

$$\begin{cases} \Delta w_i = \kappa w_i \sum_{j \neq i} w_j, \\ \Delta u_{i,1} = \kappa u_{i,1} \sum_{j \neq i} u_{j,0}. \end{cases}$$

Noting that we have $w_j < u_{j,0}$, so we can apply the comparison principle to get the claim.
   In the following, we assume that our claim is valid until $2m + 1$, that is

$$u_{i,2m} \geq w_i \geq u_{i,2m+1}.$$

Then we have the following.

Claim 3 $u_{i,2m+2} \geq w_i$.
   This can be seen by comparing the equations they satisfy:

$$\begin{cases} \Delta w_i = \kappa w_i \sum_{j \neq i} w_j, \\ \Delta u_{i,2m+2} = \kappa u_{i,2m+3} \sum_{j \neq i} u_{j,2m+1}. \end{cases}$$

By assumption, we have $u_{j,2m+1} \leq w_j$, so the claim follows from the comparison principle again.

Claim 4 $u_{i,2m+3} \leq w_i$.

This can be seen by comparing the equations they satisfy:

$$\begin{cases} \Delta w_i = \kappa w_i \sum_{j \neq i} w_j, \\ \Delta u_{i,2m+3} = \kappa u_{i,2m+3} \sum_{j \neq i} u_{j,2m+2}. \end{cases}$$

By Claim 3, we have $u_{j,2m+2} \geq w_j$, so the claim follows from the comparison principle again. $\square$

Remark 2.1.4 From our proof, we know that the uniqueness result still holds for equations of more general form:

$$\begin{cases} \Delta u_i = u_i \sum_{j \neq i} b_{ij}(x) u_j, & \text{in } \Omega \\ u_i = \varphi_i & \text{on } \partial\Omega, \end{cases}$$

where $b_{ij}(x)$ are positive (and smooth enough) functions defined in $\overline{\Omega}$, which satisfy $b_{ij} \equiv b_{ji}$.

2.2 Asymptotics in the Parabolic Case

The method in the previous section can also be used to prove the stability of solutions to the following parabolic initial-boundary value problem.

$$\begin{cases} \dfrac{\partial u_i}{\partial t} - \Delta u_i = -\kappa u_i \sum_{j \neq i} b_{ij} u_j, & \text{in } \Omega \times (0, +\infty), \\ u_i = \varphi_i, & \text{on } \partial\Omega \times (0, +\infty), \\ u_i = \phi_i, & \text{on } \Omega \times \{0\}. \end{cases} \tag{2.9}$$

Here $b_{ij} > 0$ and φ_i are those given in Theorem 2.1.1. ϕ_i are given nonnegative Lipschitz continuous functions in Ω, such that $\phi_i = \varphi_i$ on $\partial\Omega$. We prove the following theorem.

Theorem 2.2.1 *For any $\kappa \geq 0$, there exists a unique global solution U of (2.9). As $t \to +\infty$, $U(t)$ converges to the solution of (2.1) in $C(\overline{\Omega})$.*

Proof Let us consider the iteration scheme analogous to (2.3). First, we consider

$$\begin{cases} \dfrac{\partial u_{i,0}}{\partial t} - \Delta u_{i,0} = 0, & \text{in } \Omega \times (0, +\infty), \\ u_{i,0} = \varphi_i & \text{on } \partial\Omega \times (0, +\infty), \\ u_{i,0} = \phi_i & \text{on } \Omega \times \{0\}. \end{cases}$$

We know this equation has a unique positive solution $u_{i,0}(x, t)$. We also have

$$\lim_{t \to +\infty} u_{i,0}(x, t) = u_{i,0}(x),$$

where the convergence is (for example), in the space of $C^0(\overline{\Omega})$ and $u_{i,0}(x)$ is the solution of (2.2). In fact, we can prove that

$$\int_{\Omega} \left| \frac{\partial u_{i,0}}{\partial t} \right|^2 dx \le C_1 e^{-C_2 t}$$

for some positive constants C_1 and C_2.

Now the iteration can be defined as:

$$\begin{cases} \dfrac{\partial u_{i,m+1}}{\partial t} - \Delta u_{i,m+1} = -\kappa u_{i,m+1} \displaystyle\sum_{j \ne i} u_{j,m}, & \text{in } \Omega \times (0, +\infty), \\ u_{i,m+1} = \varphi_i & \text{on } \partial\Omega \times (0, +\infty), \\ u_{i,m+1} = \phi_i & \text{on } \Omega \times \{0\}. \end{cases} \qquad (2.10)$$

This is just a linear parabolic equation, and there exists a unique global solution $u_{i,m+1}(x, t)$. Differentiating (2.10) in time t, we get

$$\frac{\partial}{\partial t} \frac{\partial u_{i,m+1}}{\partial t} - \Delta \frac{\partial u_{i,m+1}}{\partial t} = -\kappa \frac{\partial u_{i,m+1}}{\partial t} \sum_{j \ne i} u_{j,m} - \kappa u_{i,m+1} \sum_{j \ne i} \frac{\partial u_{j,m}}{\partial t}. \qquad (2.11)$$

By the induction assumption and maximum principle, we know there exist constants C_m', $C_{m,1}$ and $C_{m,2}$ such that for $t > 1$,

$$\sum_{j \ne i} u_{j,m+1} \le C_m', \qquad (2.12)$$

$$\int_{\Omega} \left| \frac{\partial u_{i,m}}{\partial t} \right|^2 dx \le C_{m,1} e^{-C_{m,2} t}. \qquad (2.13)$$

Multiplying (2.11) by $\frac{\partial u_{i,m+1}}{\partial t}$, with the help of (2.12), we get (note that we have the boundary condition $\frac{\partial u_{i,m+1}}{\partial t} = 0$ on $\partial\Omega$)

$$\frac{d}{dt} \int_{\Omega} \frac{1}{2} \left| \frac{\partial u_{i,m+1}}{\partial t} \right|^2 + \int_{\Omega} \left| \nabla \frac{\partial u_{i,m+1}}{\partial t} \right|^2 \le \kappa C_m' \int_{\Omega} \sum_{j \ne i} \left| \frac{\partial u_{j,m}}{\partial t} \right| \left| \frac{\partial u_{i,m+1}}{\partial t} \right|.$$

Using Cauchy inequality, we get

$$\frac{d}{dt}\int_\Omega \frac{1}{2}\left|\frac{\partial u_{i,m+1}}{\partial t}\right|^2 + \int_\Omega \left|\nabla \frac{\partial u_{i,m+1}}{\partial t}\right|^2$$

$$\leq \kappa C'_m \left(\int_\Omega \sum_{j\neq i}\left|\frac{\partial u_{j,m}}{\partial t}\right|^2\right)^{\frac{1}{2}}\left(\int_\Omega \left|\frac{\partial u_{i,m+1}}{\partial t}\right|^2\right)^{\frac{1}{2}}.$$

By (2.13) and the Poincare inequality, we get

$$\int_\Omega \left|\frac{\partial u_{i,m+1}}{\partial t}\right|^2 dx \leq C_{m+1,1}e^{-C_{m+1,2}t},$$

for some positive constants $C_{m+1,1}$ and $C_{m+1,2}$.

By standard parabolic estimate, this also imply

$$\sup_\Omega \left|\frac{\partial u_{i,m+1}}{\partial t}\right| \leq C_{m+1,1}e^{-C_{m+1.2}t},$$

for another two constants $C_{m+1,1}$ and $C_{m+1,2}$. This implies

$$\lim_{t\to+\infty} u_{i,m+1}(x,t) = u_{i,m+1}(x),$$

where $u_{i,m+1}(x)$ is the solution of (2.3). Furthermore, the convergence can be taken (for example) in the space of $C^0(\overline{\Omega})$.

The same method of Sect. 2.2 gives, in $\Omega \times (0,+\infty)$

$$u_{i,0} > \cdots > u_{i,2m} > u_{i,2m+2} > \cdots > u_i > \cdots > u_{i,2m+1} > u_{i,2m-1} > \cdots > u_{i,1}.$$

Now our Theorem 2.2.1 can be easily seen. In fact, $\forall \varepsilon > 0$, there exists a m, such that

$$\max_{\overline{\Omega}}\left|u_{i,2m}(x) - u_i(x)\right| < \varepsilon$$

and

$$\max_{\overline{\Omega}}\left|u_{i,2m+1}(x) - u_i(x)\right| < \varepsilon.$$

We also have that there exists a $T > 0$, depending on m only, such that, $\forall t > T$,

$$\max_{\overline{\Omega}}\left|u_{i,2m}(x,t) - u_{i,2m}(x)\right| < \varepsilon,$$

and

$$\max_{\overline{\Omega}}\left|u_{i,2m+1}(x,t) - u_{i,2m+1}(x)\right| < \varepsilon.$$

Combing these together, we get $\forall t > T$,

$$\max_{\overline{\Omega}} \left| u_i(x, t) - u_i(x) \right| < 4\varepsilon.$$

This implies that $u_i(x, t)$ converges to the solution $u_i(x)$ of (2.1) as $t \longrightarrow +\infty$, uniformly on $\overline{\Omega}$. (If the boundary values are sufficiently smooth, the convergence in Theorem 2.2.1 can be improved to be smooth enough.)   □

2.3 A Uniform Lipschitz Estimate

Finally, by the same idea as in the previous sections, we prove the uniform Lipschitz estimates for solutions to the above two problems (2.9) and (2.2).

Theorem 2.3.1 *There exists a constant $C > 0$ independent of κ, such that for any $\kappa \geq 0$ and solution $(u_{i,\kappa})$ of (2.1), we have*

$$\sup_{\Omega} |\nabla u_{i,\kappa}| \leq C.$$

Theorem 2.3.2 *There exists a constant $C > 0$ independent of κ, such that for any $\kappa \geq 0$ and solution $(u_{i,\kappa})$ of (2.9), we have*

$$\sup_{\Omega \times [0,+\infty)} Lip(u_{i,\kappa}) \leq C.$$

We will only treat the parabolic case. The elliptic case is similar.

We need an additional assumption on the initial-boundary values here. Let Φ_i be the solution of

$$\begin{cases} \dfrac{\partial \Phi_i}{\partial t} - \Delta \Phi_i = 0, & \text{in } \Omega \times (0, +\infty), \\[2mm] \Phi_i = \varphi_i, & \text{on } \partial\Omega \times (0, +\infty), \\[2mm] \Phi_i = \phi_i, & \text{on } \Omega \times \{0\}. \end{cases} \tag{2.14}$$

We assume that Φ_i are Lipschitz continuous on the closure of $\Omega \times (0, +\infty)$. Note that by comparison principle, we have (see [11] for the proof in the elliptic case)

$$\begin{cases} \Phi_i \geq u_{i,\kappa}, \\[2mm] \Phi_i - \sum_{j \neq i} \Phi_j \leq u_{i,\kappa} - \sum_{j \neq i} u_{j,\kappa}. \end{cases} \tag{2.15}$$

First differentiating (2.9) in a space direction e we obtain an equation for $D_e u := e \cdot \nabla u$:

$$\left(\frac{\partial}{\partial t} - \Delta \right) D_e u_{i,\kappa} = -\kappa D_e u_{i,\kappa} \sum_{j \neq i} u_{j,\kappa} - \kappa u_{i,\kappa} \sum_{j \neq i} D_e u_{j,\kappa}.$$

Now using the Kato inequality for smooth functions ϕ

$$|\nabla|\phi|| = |\nabla\phi| \quad \text{a.e.,} \quad |\Delta|\phi|| \geq |\Delta\phi|,$$

we have

$$\left(\frac{\partial}{\partial t} - \Delta\right)|D_e u_{i,\kappa}| \leq -\kappa|D_e u_{i,\kappa}| \sum_{j\neq i} u_{j,\kappa} + \kappa u_{i,\kappa} \sum_{j\neq i} |D_e u_{j,\kappa}|.$$

Summing these in i, we get

$$\left(\frac{\partial}{\partial t} - \Delta\right) \sum_i |D_e u_{i,\kappa}| \leq 0.$$

By the assumption on Φ_i and (2.15), we have

$$\sup_{\partial\Omega\times(0,+\infty)} \left|\frac{\partial u_{i,\kappa}}{\partial \nu}\right| \leq C,$$

for all i, where ν is the outward unit normal vector and C is independent of κ. With the assumption of Lipschitz continuity of the boundary values on $\partial\Omega \times (0,+\infty)$, we in fact have

$$\sup_{\partial\Omega\times(0,+\infty)} |\nabla u_{i,\kappa}| \leq C,$$

with a constant C independent of κ again. Next, we also have at $t = 0$, $u_{i,\kappa} = \phi_i$, so

$$\sup_{\Omega\times\{0\}} |\nabla u_{i,\kappa}| = \sup_{\Omega} |\nabla\phi_i|.$$

Now the maximum principle implies a global uniform bound:

$$\sup_{\Omega\times[0,+\infty)} |\nabla u_{i,\kappa}| \leq C.$$

Then by a standard method we can get the uniform Lipschitz bound with respect to the parabolic distance.

Remark 2.3.3 Without the boundary regularity, we can still get an interior uniform bound. Multiplying the equation by $u_{i,\kappa}$ and integrating by parts, we can get a L^2 bound for any $T > 0$

$$\sum_i \int_T^{T+1} \int_\Omega |\nabla u_{i,\kappa}|^2 \leq C,$$

with C independent of κ and T. Then we can use the mean value property for sub-caloric (or subharmonic function) to give a uniform upper bound of $|\nabla u_{i,\kappa}|$.

Remark 2.3.4 If we consider the original Lotka–Volterra system

$$\frac{\partial u_i}{\partial t} - \Delta u_i = a_i u_i - u_i^2 - \kappa u_i \sum_{j \neq i} u_j,$$

with homogeneous Dirichlet boundary condition, the above results still hold. In fact, we only need to prove a boundary gradient estimate, which can be guaranteed by the following argument: if we define v_i to be the solution of

$$\frac{\partial v_i}{\partial t} - \Delta v_i = a_i v_i - v_i^2,$$

with the same initial value, then by the maximum principle we have for each κ

$$u_{i,\kappa} \leq v_i,$$

which, together with the boundary condition, implies

$$\left| \frac{\partial u_{i,\kappa}}{\partial \nu} \right| \leq \left| \frac{\partial v_i}{\partial \nu} \right|,$$

where ν is the unit outward normal vector to $\partial \Omega$; using the boundary condition once again we get on the boundary

$$|\nabla u_{i,\kappa}| \leq |\nabla v_i|,$$

where the right hand side is independent of κ.

Chapter 3
Uniqueness in the Singular Limit

Abstract In this chapter we study some properties of solutions to the problem (1.5). We first give an interior Hausdorff measure estimate of the free boundary of these solutions. Then we prove the uniqueness of the solution to (1.5), which then can be identified as the minimizer of a functional. Finally, we prove the uniqueness of solutions to the corresponding parabolic problem of (1.5).

3.1 An Interior Measure Estimate

In this section, we study some properties of solutions to the following problem.

$$
\begin{cases}
\Delta u_i \geq 0, & \text{in } \Omega, \\
\Delta\left(u_i - \sum_{j \neq i} u_j\right) \leq 0, & \text{in } \Omega, \\
u_i = \varphi_i, & \text{on } \partial\Omega, \\
u_i u_j = 0, & \text{in } \Omega.
\end{cases}
\tag{3.1}
$$

φ_i are given nonnegative Lipschitz continuous functions on $\partial\Omega$, satisfying

$$\varphi_i \varphi_j = 0 \quad \text{if } i \neq j.$$

As mentioned in the introduction, we assume $u_i \in H^1(\Omega)$, and the above two inequalities are understood in the H^1 sense. Note that u_i can be proved to be continuous, see [9]. We first give an interior measure estimate of free boundaries. For any $N > 0$, define

$$
H_N^1 := \Big\{ u : B_1(0) \to \Sigma, \text{ satisfying (3.1) except the boundary conditions, and}
$$

$$
\frac{\int_{B_1(0)} \sum_i |\nabla u_i^2|}{\int_{\partial B_1(0)} \sum_i u_i^2} \leq N, \int_{\partial B_{\frac{1}{2}}(0)} \sum_i u_i^2 = 1 \Big\}.
$$

The main result of this section is the following theorem.

K. Wang, *Free Boundary Problems and Asymptotic Behavior of Singularly Perturbed Partial Differential Equations*, Springer Theses,
DOI 10.1007/978-3-642-33696-6_3, © Springer-Verlag Berlin Heidelberg 2013

Theorem 3.1.1 *There exists a constant $C(N,n)$ depending only on N and the dimension n, such that for any $U \in H_N^1$,*

$$H^{n-1}\left(B_{\frac{1}{2}}(0) \cap \mathscr{F}(u)\right) \le C(N,n).$$

Here H^{n-1} denotes the $n-1$ dimensional Hausdorff measure.

Throughout this section, N will be a fixed positive constant.

In order to prove this theorem, we need some lemmas (following the same ideas in Sect. 5.2 of [26]). The first one is a compactness result for functions in H_N^1 with a given $N > 0$.

Lemma 3.1.2 H_N^1 *is compact in* $L^2(B_1(0))$.

Proof First from the Almgren monotonicity formula, we have a well-known doubling property (cf. Sect. 2 in [4]), which implies

$$\int_{\partial B_1(0)} \sum_i u_i^2 \le C(N),$$

where $C(N)$ depends only on N. By the definition of H_N^1 and the Poincare inequality (noting here we have a bound on the boundary integral), we have

$$\int_{B_1(0)} \sum_i |\nabla u_i^2| + \sum_i u_i^2 \le C(N),$$

for another constant $C(N)$. So for any sequence $u_m \in H_N^1$, there exists a subsequence converging to u, weakly in $H^1(B_1)$ and strongly in $L^2(B_1)$.

We claim that the limit u lies in H_N^1. First, we know those properties in (3.1) are preserved under weak convergence in $H^1(B_1)$ and strong convergence in $L^2(B_1)$. Next, we claim for any $r < 1$, u_m converges to u strongly in $H^1(B_r)$. This is because, if we take a smooth cut-off function ζ, from the continuity of u_m and the fact that $\Delta u_{i,m}$ is a Radon measure supported on $\partial\{u_m > 0\}$, we have

$$0 = \int \Delta u_{i,m} \cdot u_{i,m} \zeta^2 = -\int |\nabla u_{i,m}|^2 \zeta^2 + 2\zeta u_{i,m} \nabla u_{i,m} \nabla \zeta.$$

So from the weak convergence of $u_{i,m}$ in $H^1(B_1)$ and the fact that $u_{i,m}$ converges to u_i uniformly, we get

$$\lim_{m \to +\infty} \int |\nabla u_{i,m}|^2 \zeta^2 = \int |\nabla u_i|^2 \zeta^2.$$

From Trace Theorem in H^1, we also have

$$\int_{\partial B_{\frac{1}{2}}(0)} \sum_i u_i^2 = \lim_{m \to +\infty} \int_{\partial B_{\frac{1}{2}}(0)} \sum_i u_{i,m}^2 = 1.$$

Thus for any $r < 1$

$$\frac{\int_{B_r(0)} \sum_i |\nabla u_i^2|}{\int_{\partial B_r(0)} \sum_i u_i^2} = \lim_{m \to +\infty} \frac{\int_{B_r(0)} \sum_i |\nabla u_{i,m}^2|}{\int_{\partial B_r(0)} \sum_i u_{i,m}^2} \leq N.$$

By the monotonicity and continuity of the frequency, this implies

$$\frac{\int_{B_1(0)} \sum_i |\nabla u_i^2|}{\int_{\partial B_1(0)} \sum_i u_i^2} \leq N. \qquad \square$$

The next step is to divide the free boundary into two parts: the good parts are those which are uniformly smooth (the gradient has a uniform lower bound there), while for the bad parts we have a control on its size. In the following, we shall denote the free boundary of u as $\mathscr{F}(u)$.

Lemma 3.1.3 *For any $u \in H_N^1$, there exist finite balls $B_{r_k}(x_k)$ with $r_k \leq \frac{1}{2}$ such that*

$$\left\{ x \in B_{\frac{1}{2}}, \sum_i |\nabla u_i^2| \leq \gamma(N) \right\} \cap \mathscr{F}(u) \subset \bigcup_k B_{r_k}(x_k),$$

and

$$\sum_k r_k^{n-1} \leq \frac{1}{2},$$

where $\gamma(N)$ is a constant depending only on the dimension n and N.

Proof For any $u_0 \in H_N^1$, the singular set of the free boundary $sing(\mathscr{F}(u_0))$ has vanishing $n-1$ dimensional Hausdorff measure:

$$H^{n-1}\big(sing(\mathscr{F}(u_0))\big) = 0.$$

So there exist finitely many balls $B_{r_k}(x_k)$ with $r_k \leq \frac{1}{2}$ such that

$$sing(\mathscr{F}(u_0)) \subset \bigcup_k B_{\frac{r_k}{2}}(x_k)$$

and

$$\sum_k r_k^{n-1} \leq \frac{1}{2^n}.$$

Of course, there exists a constant $\gamma(u_0) > 0$, such that, on the set $B_{\frac{1}{2}} \cap \mathscr{F}(u_0) \setminus (\bigcup_k B_{\frac{r_k}{2}}(x_k))$,

$$\sum_i |\nabla u_{i,0}|^2 \geq 3\gamma(u_0). \qquad (3.2)$$

Now we claim there exists a $\varepsilon(u_0) > 0$ such that for any $u \in H_N^1$ with $\|u - u_0\|_{L^2(B_1)} \le \varepsilon(u_0)$, on $B_{\frac{1}{2}} \cap \mathscr{F}(u) \setminus (\bigcup_k B_{r_k}(x_k))$,

$$\sum_i |\nabla u_i|^2 \ge \gamma(u_0).$$

With the compactness of H_N^1 in $L^2(B_1)$, our conclusion can be easily seen.

Assume this claim is not true, then there exists a sequence of $u_m \in H_N^1$ with $\|u_m - u_0\|_{L^2(B_1)} \le \frac{1}{m}$, but $\exists x_m \in B_{\frac{1}{2}} \cap \mathscr{F}(u_m) \setminus (\bigcup_k B_{r_k}(x_k))$,

$$\sum_i |\nabla u_{i,m}|^2(x_m) \le \gamma(u_0). \tag{3.3}$$

Then from the uniform Lipschitz estimate (cf. Corollary 3.2 in [4]), we know that u_m converge to u uniformly on any compact subset of $B_1(0)$. This implies for any $\delta > 0$, for m large enough depending only on δ, $\mathscr{F}(u_m)$ is in the δ neighborhood of $\mathscr{F}(u)$. However, near $B_{\frac{1}{2}} \cap \mathscr{F}(u_0) \setminus (\bigcup_k B_{r_k}(x_k))$, locally, there exist exactly two components of u_0 which are non-vanishing here, without loss of generality, assuming to be $u_{1,0}$ and $u_{2,0}$, which satisfies

$$\Delta(u_{1,0} - u_{2,0}) = 0.$$

Using the same method of the Clean Up lemma in [2], we can show that locally, for m large, we also have, only $u_{1,m}$ and $u_{2,m}$ are non-vanishing (this can also be proven by the upper semicontinuity of the frequency function under the convergence of $u_m \to u$ and $x_m \to x$). Then we also have locally

$$\Delta(u_{1,m} - u_{2,m}) = 0.$$

In view of their convergence in L_{loc}^2, we have here locally

$$u_{1,m} - u_{2,m} \to u_{1,0} - u_{2,0} \quad \text{smoothly.}$$

Now coming back to (3.3), without loss of generality, assuming $x_m \to x_0$, which lies in $B_{\frac{1}{2}} \cap \mathscr{F}(u_0) \setminus (\bigcup_k B_{r_k}(x_k))$, we can take the limit in (3.3) to get

$$\sum_i |\nabla u_{i,0}|^2(x_0) \le \gamma(u_0),$$

which contradicts (3.2). $\square$

First, we need to control the measure of the good part. This needs a comparison with some standard models, for example, in [26], they use the comparison with harmonic functions. But here we have no such smooth model to compare. Instead, we will compare it with the homogeneous elements in H_N^1, which has the property

$$u(rx) = r^d u(x), \quad \text{for some } d > 0.$$

It can be represented by $u(r\theta) = r^d \varphi(\theta)$, for φ defined on $\mathbb{S}^{n-1}$, satisfying

$$
\begin{cases}
(\Delta_\theta + \lambda)\varphi_i \geq 0, \\[2mm]
(\Delta_\theta + \lambda)\left(\varphi_i - \sum_{j \neq i} \varphi_j\right) \leq 0, \\[2mm]
\varphi_i \geq 0, \\[2mm]
\varphi_i \varphi_j = 0,
\end{cases}
\tag{3.4}
$$

where λ satisfies $d(d+n-2) = \lambda$ and $\lambda \leq N$, and Δ_θ is the Laplacian on $\mathbb{S}^{n-1}$. By induction on the dimension, we can assume

$$
H^{n-2}\big(\mathscr{F}(\varphi) \cap \mathbb{S}^{n-1}\big) \leq C(N, n).
$$

Note here in dimension $n = 2$, each φ can be written explicitly.

Lemma 3.1.4 *With the assumptions of the preceding lemma, if moreover*

$$
\frac{\int_{B_1(0)} \sum_i |\nabla u_i^2|}{\int_{\partial B_1(0)} \sum_i u_i^2} - N(u, 0) \leq \sigma,
$$

where σ is a constant depending only on the dimension n and N, and

$$
N(u, 0) = \lim_{r \to 0} \frac{r \int_{B_r(0)} \sum_i |\nabla u_i^2|}{\int_{\partial B_r(0)} \sum_i u_i^2}.
$$

Then

$$
H^{n-1}\left(\mathscr{F}(u) \cap B_{\frac{1}{2}} \Big\backslash \left(\bigcup_k B_{r_k}(x_k)\right)\right) \leq C(N),
$$

where $C(N)$ is a constant depending only on the dimension n and N.

Proof Take a $\delta > 0$ small enough. If we choose σ small enough too, then by compactness there exists a homogeneous $w \in H_N^1$ such that

$$
\|u - w\|_{L^2(B_1)} \leq \delta,
$$

and $\mathscr{F}(u)$ is in the δ neighborhood of $\mathscr{F}(w)$. Define

$$
S_1 := \left\{ x \in B_{\frac{1}{2}}, \sum_i |\nabla u_i|^2 \geq \gamma(N) \right\} \cap \mathscr{F}(u),
$$

$$
S_2 := \left\{ x \in B_{\frac{1}{2}}, \sum_i |\nabla w_i|^2 \geq \gamma(N) \right\} \cap \mathscr{F}(w).
$$

If σ is small, S_1 is in the δ neighborhood of S_2, too. Take a $\varepsilon \gg \delta$, and take a maximal ε separated sets $\{y_k\}$ of S_2. Then we have $dist(y_k, y_l) \geq \frac{\varepsilon}{2}$ and $S_2 \subset \bigcup_k B_\varepsilon(y_k)$.

In each $B_\varepsilon(y_k)$, w has exactly two components which are non-vanishing. Moreover, the free boundary $\mathscr{F}(w) \cap B_\varepsilon(y_k)$ can be represented by the graph of a C^1 function defined on the tangent plane to $\mathscr{F}(w)$ at y_k. Now if δ is small enough, this property is also valid for u. The same method of Lemma 5.25 in [26] gives our conclusion. □

Now the proof of Theorem 3.1.1 can be easily done by an iteration procedure exactly as in [26]. Here, we just need to note that in Lemma 3.1.3, those radius r_i can be chosen arbitrarily small so that the assumptions in Lemma 3.1.4 are satisfied.

Finally, we give a theorem on the uniform estimate of the measure of the level surface $\{u_i = \delta\}$.

Theorem 3.1.5 *For $u \in H_N^1$, $\forall \delta > 0$ and $1 \leq i \leq M$, we have*

$$H^{n-1}\left(B_{\frac{1}{2}} \cap \{u_i = \delta\}\right) \leq C(N).$$

This is also valid for the local energy minimizing map.

Proof First, because each u_i is subharmonic, from the $L^2(B_1)$ bound we have

$$\sup_{B_{\frac{1}{2}}} u_i \leq C(N). \tag{3.5}$$

We claim that, $\forall \delta > 0$, $\exists C(\delta, N)$, such that, $\forall t > \delta$

$$H^{n-1}\left(B_{\frac{1}{2}} \cap \{u_i = t\}\right) \leq C(\delta, N). \tag{3.6}$$

If this is not true, then $\exists t_k \geq \delta$ and $u_k \in H_N^1$ such that

$$H^{n-1}\left(B_{\frac{1}{2}} \cap \{u_{i,k} = t_k\}\right) \geq k.$$

By (3.5), we can assume $t_k \to t \geq \delta$. By the compactness of H_N^1, we can assume $u_k \to u$ in $L^2(B_1)$. By the uniform Hölder continuity [2], we can also assume $u_{i,k} \to u_i$ in $C(B_{\frac{2}{3}})$.

If $u_i \equiv 0$, then for k large

$$\sup_{B_{\frac{2}{3}}} u_{i,k} < \frac{t}{2}.$$

This is impossible, so u_i is not 0. In fact, $\{u_i = t\} \cap B_{\frac{1}{2}} \neq \emptyset$. Because in $\{u_i > 0\}$, u_i is harmonic, we have

$$H^{n-1}\left(B_{\frac{1}{2}} \cap \{u_i = t\}\right) < +\infty.$$

We also have for k large, $B_{\frac{1}{2}} \cap \{u_{i,k} = t_k\}$ lies in a small neighborhood of $B_{\frac{1}{2}} \cap \{u_i = t\}$. Note that these level sets are smooth hypersurfaces outside a closed set of Hausdorff dimension $\leq n - 2$. In this neighborhood, $u_{i,k}$ converge to u_i smoothly, so for k large, $\exists C > 0$ such that

$$H^{n-1}\left(B_{\frac{1}{2}} \cap \{u_{i,k} = t_k\}\right) \leq C.$$

This is a contradiction, so our claim follows.

Now we can use an iteration to prove our theorem. For any $u \in H_N^1$, take a covering of the singular set of the free boundary as in Lemma 3.1.3:

$$sing(\mathscr{F}(u)) \subset \bigcup_k B_{\frac{r_k}{2}}(x_k)$$

with

$$\sum_k r_k^{n-1} \leq \frac{1}{2^n}.$$

By the uniform interior Lipschitz estimate, $\exists \delta(N) > 0$, such that

$$\sup_{\bigcup_k B_{r_k}(x_k)} \sum_i u_i \leq \delta(N).$$

While for $x \in B_{\frac{1}{2}} \cap \mathscr{F}(u) \setminus \bigcup_k B_{\frac{r_k}{2}}(x_k)$,

$$\sum_i |\nabla u_i|^2 \geq \gamma(N),$$

so $\exists C(N)$, such that $\forall \delta > 0$

$$H^{n-1}\left(B_{\frac{1}{2}} \cap \{u_i = \delta\} \setminus \bigcup_k B_{\frac{r_k}{2}}(x_k)\right) \leq C(N).$$

Now we can rescale u in B_{r_k}:

$$\hat{u} = L_k u(x_k + r_k x).$$

If we choose L_k appropriately, $\hat{u}$ is still in H_N^1, and we can iterate the above procedure. This iteration will stop in finitely many times and at last we get our original estimate. $\qquad \square$

3.2 Uniqueness in the Singular Limit

In this section we prove the uniqueness of solutions to Eq. (3.1).

Theorem 3.2.1 *Given boundary conditions φ as above, there exists a unique solution of (3.1). Moreover, this solution is the minimizer of the energy functional (1.6).*

In order to prove the energy minimizing property, we need to prove that for any given Lipschitz map $w : \Omega \to \Sigma$ such that $w \equiv u$ outside a compact set $\Omega' \subset\subset \Omega$, we have

$$\int_{\Omega'} \sum_i |\nabla u_i|^2 \, dx \le \int_{\Omega'} \sum_i |\nabla w_i|^2 \, dx.$$

In fact, we will prove that, if v minimizes $\{\int_{\Omega'} \sum_i |\nabla w_i|^2 \, dx : w \equiv u \text{ outside } \Omega'\}$, then $u \equiv v$ in Ω'.

First, $\forall i$, in $\{u_i > 0\}$ (or $\{v_i > 0\}$), u_i (or v_i) is harmonic, thus real analytic. If we enlarge Ω', we can assume $\partial \Omega'$ is real analytic and smooth (without singularity). Choose two constant $\delta > \sigma > 0$, such that the level surfaces $\{u_i = \delta\}$ and $\{v_i = \delta\}$, $\forall i$, are regular real analytic hypersurface up to the boundary (that is, $\{u_i = \delta\} \cap \partial \Omega'$ and $\{v_i = \delta\} \cap \partial \Omega'$ are regular real analytic hypersurface in $\partial \Omega'$). With this setting, we know that the divergence theorem is valid for domains separated by these hypersurfaces.

Define $u_i^\delta := \max\{u_i - \delta, 0\}$ and $v_i^\sigma := \max\{v_i - \sigma, 0\}$. We consider the geodesic homotopy $u^t : \Omega \to \Sigma$ between u^δ and v^σ for $t \in [0, 1]$, that is, $u^t(x)$ is the point on the unique geodesic between $u^\delta(x)$ and $v^\sigma(x)$ which is characterized uniquely by $d(u^t(x), u^\delta(x)) = t d(u^\delta(x), v^\sigma(x))$ (here d denotes the intrinsic distance of Σ).

We can write down the expression of $u^t(x)$ explicitly from the concrete form of Σ, following the construction of the test functions used in [10]. In the set $A_i := \{x : u_i^\delta(x) > 0, v_i^\sigma(x) > 0\}$:

$$u_i^t(x) = (1 - t)u_i^\delta(x) + t v_i^\sigma(x); \tag{3.7}$$

in the set $B_{ij} := \{x : u_i^\delta(x) > 0, v_j^\sigma(x) > 0 \text{ and } u_i^\delta(x) - t(u_i^\delta(x) + v_j^\sigma(x)) > 0\}$ for some $j \ne i$:

$$u_i^t(x) = u_i^\delta(x) - t\left(u_i^\delta(x) + v_j^\sigma(x)\right); \tag{3.8}$$

in the set $C_{ij} := \{x : u_j^\delta(x) > 0, v_i^\sigma(x) > 0 \text{ and } t(u_j^\delta(x) + v_i^\sigma(x)) - u_j^\delta(x) > 0\}$ for some $j \ne i$:

$$u_i^t(x) = t\left(u_j^\delta(x) + v_i^\sigma(x)\right) - u_j^\delta(x); \tag{3.9}$$

in the set $D_i := \{x : u_i^\delta(x) > 0, v_j^\sigma(x) = 0, \forall j\}$:

$$u_i^t(x) = (1 - t)u_i^\delta(x); \tag{3.10}$$

in the set $E_i := \{x : v_i^\sigma(x) > 0, u_j^\delta(x) = 0, \forall j\}$:

$$u_i^t(x) = t v_i^\sigma(x); \tag{3.11}$$

in the remaining part $u_i^t(x) \equiv 0$ for each i.

Now we have (note that in $\{u_i^\delta(x) > 0\}$, $\nabla u_i^\delta(x) = \nabla u_i(x)$ a.e.)

$$\int_\Omega |\nabla u_i^t(x)|^2\, dx = \int_{A_i} |(1-t)\nabla u_i(x) + t\nabla v_i(x)|^2\, dx$$

$$+ \sum_{j\neq i} \int_{B_{ij}} |(1-t)\nabla u_i(x) - t\nabla v_j(x)|^2\, dx$$

$$+ \sum_{j\neq i} \int_{C_{ij}} |-(1-t)\nabla u_j(x) + t\nabla v_i(x)|^2\, dx$$

$$+ \int_{D_i} |(1-t)\nabla u_i(x)|^2\, dx$$

$$+ \int_{E_i} |t\nabla v_i(x)|^2\, dx. \tag{3.12}$$

We need to compute $\frac{dE}{dt}|_{t=0}$. Noticing that in the first term of (3.12) the domain is fixed, and $\Delta u_i = 0$ in the open set $\{u_i > \delta\}$, we can integrate by parts to get

$$2\int_{A_i} \nabla u_i(x)\big(\nabla v_i(x) - \nabla u_i(x)\big)\, dx = 2\int_{\partial A_i} \frac{\partial u_i}{\partial v}\big[v_i(x) - u_i(x)\big]. \tag{3.13}$$

Now we have

$$\partial A_i = A_{i,1} \cup A_{i,2} \cup A_{i,3},$$

where

$$\begin{cases} A_{i,1} = \{u_i = \delta\} \cap \{v_i > \sigma\}, \\ A_{i,2} = \{v_i = \sigma\} \cap \{u_i > \delta\}, \\ A_{i,3} = \partial\Omega' \cap \{v_i > \sigma\} \cap \{u_i > \delta\}, \\ A_{i,4} = \{u_i = \delta, v_i = \sigma\}. \end{cases}$$

On $A_{i,3}$ we have $u_i = v_i$, while the $n-1$ dimension Hausdorff measure of $A_{i,4}$ is 0 (it lies in the interior of Ω'), so they do not appear in (3.13). On $A_{i,1}$ we have $u_i = \delta$ and on $A_{i,2}$ we have $v_i = \sigma$. Thus, we get

$$2\int_{A_{i,1}} \frac{\partial u_i}{\partial v_{i,1}^-} v_i - 2\int_{A_{i,2}} \frac{\partial u_i}{\partial v_{i,2}^-} u_i - 2\delta \int_{A_{i,1}} \frac{\partial u_i}{\partial v_{i,1}^-} + 2\sigma \int_{A_{i,2}} \frac{\partial u_i}{\partial v_{i,2}^-}. \tag{3.14}$$

Here $v_{i,1}^-$ and $v_{i,2}^-$ are the outward unit normal vector to $\partial\{u_i > \delta\}$ and $\partial\{v_i > \delta\}$, respectively.

Next, let us consider the second term in (3.12). Here we must be careful because the domain changes as t changes. That is

$$B_{ij} := \big\{x : u_i(x) > \delta,\, v_j(x) > \sigma\big\} \setminus \Big\{u_i^\delta(x) < \frac{t}{1-t} v_j^\sigma(x)\Big\}.$$

So the second term of (3.12) can be written as

$$\sum_{j \neq i} \int_{\{x : u_i(x) > \delta, v_j(x) > \sigma\}} \left| (1 - t) \nabla u_i(x) - t \nabla v_j(x) \right|^2 dx$$

$$- \sum_{j \neq i} \int_{\{x : u_i^\delta(x) \leq \frac{t}{1-t} v_j^\sigma(x)\}} \left| (1 - t) \nabla u_i(x) - t \nabla v_j(x) \right|^2 dx.$$

In the first term, the domain is fixed, and the derivative can be calculated directly. The second term can be written in another form using the Co-Area formula (see, for example, [35]):

$$\sum_{j \neq i} \int_0^{\frac{t}{1-t}} \left[\int_{\{x : \frac{u_i^\delta}{v_j^\sigma} = \tau\}} \frac{|(1 - t) \nabla u_i(x) - t \nabla v_j(x)|^2}{|\nabla \frac{u_i^\delta}{v_j^\sigma}|} \right] d\tau.$$

Its derivative at $t = 0$ is

$$\sum_{j \neq i} \int_{\{x : \frac{u_i^\delta}{v_j^\sigma} = 0\}} \frac{|\nabla u_i(x)|^2}{|\nabla \frac{u_i^\delta}{v_j^\sigma}|},$$

After calculation, we get

$$\sum_{j \neq i} \int_{\{x : u_i = \delta, v_j > \sigma\}} |\nabla u_i(x)| v_j^\sigma(x).$$

In conclusion, we get that the derivative of the second term of (3.12) at $t = 0$ is

$$-2 \int_{\{u_i > \delta, v_j > \sigma\}} \nabla u_i(x) \left(\nabla v_j(x) + \nabla u_i(x) \right) dx - \int_{\{u_i = \delta\} \cap \{v_j > \sigma\}} |\nabla u_i(x)| v_j.$$

$$(3.15)$$

Through an integration by parts (and the same remark as before concerning this procedure of integration by parts), the first term of (3.15) can be transformed into the boundary term, and notice that on $\{u_i = \delta\}$ we have $u_i = \delta$ and on $\{v_j = \sigma\}$ we have $v_j = \sigma$, so the first term of (3.15) is

$$-2 \int_{\{u_i = \delta\} \cap \{v_j > \sigma\}} \frac{\partial u_i}{\partial v_{i,1}^-} v_j - 2 \int_{\{v_j = \sigma\} \cap \{u_i > \delta\}} \frac{\partial u_i}{\partial v_{i,2}^-} u_i$$

$$- 2\delta \int_{\{u_i = \delta\} \cap \{v_j > \sigma\}} \frac{\partial u_i}{\partial v_{i,1}^-} - 2\sigma \int_{\{v_j = \sigma\} \cap \{u_i > \delta\}} \frac{\partial u_i}{\partial v_{i,2}^-}.$$

Here $v_{i,1}^-$ and $v_{i,2}^-$ are the outward normal vectors to $\partial \{u_i > \delta\}$ and $\partial \{v_j > \delta\}$, respectively.

The third term can be calculated similarly (notice that the domain is enlarged so here the positive sign comes out):

$$\int_{\{u_j=\delta\}\cap\{v_i>\sigma\}} |\nabla u_j(x)| v_i^\sigma. \tag{3.16}$$

The result from the fourth term is

$$
\begin{aligned}
\frac{d}{dt}\int_{E_i}(1-t)^2|\nabla u_i|^2 &= -2\int_{\{u_i>\delta\}\cap(\bigcup_j\{v_j<\sigma\})}|\nabla u_i|^2 \\
&= -2\int_{\{u_i=\delta\}\cap(\bigcup_j\{v_j<\sigma\})}\frac{\partial u_i}{\partial v_{i,1}^-}u_i \\
&\quad -2\sum_j\int_{\{u_i>\delta\}\cap\{v_j=\sigma\}}\frac{\partial u_i}{\partial v_{i,2}^+}u_i,
\end{aligned}
$$

where $v_{i,1}^-$ and $v_{i,2}^+$ are the outward normal vectors to $\partial\{u_i>\delta\}$ and $\partial\{v_i<\delta\}$, respectively.

The fifth term is of the order t^2, so there is no contribution to $\frac{dE}{dt}|_{t=0}$. Now put all of these terms for all i together:

$$
\begin{aligned}
\frac{dE}{dt}\bigg|_{t=0} &= 2\sum_i\int_{\{u_i=\delta\}\cap\{v_i>\sigma\}}\frac{\partial u_i}{\partial v_{i,1}^-}v_i - 2\sum_i\int_{\{v_i=\sigma\}\cap\{u_i>\delta\}}\frac{\partial u_i}{\partial v_{i,2}^-}u_i \\
&\quad -2\sum_i\delta\int_{\{u_i=\delta\}\cap\{v_i>\sigma\}}\frac{\partial u_i}{\partial v_{i,1}^-} + 2\sum_i\sigma\int_{\{v_i=\sigma\}\cap\{u_i>\delta\}}\frac{\partial u_i}{\partial v_{i,2}^-} \\
&\quad -\sum_{i\neq j}\int_{\{u_i=\delta\}\cap\{v_j>\sigma\}}|\nabla u_i(x)|v_j^\sigma \\
&\quad -2\sum_{i\neq j}\int_{\{u_i=\delta\}\cap\{v_j>\sigma\}}\frac{\partial u_i}{\partial v_{i,1}^-}v_j - 2\sum_{i\neq j}\int_{\{v_j=\sigma\}\cap\{u_i>\delta\}}\frac{\partial u_i}{\partial v_{i,2}^-}u_i \\
&\quad -2\sum_{i\neq j}\delta\int_{\{u_i=\delta\}\cap\{v_j>\sigma\}}\frac{\partial u_i}{\partial v_{i,1}^-} - 2\sum_{i\neq j}\sigma\int_{\{v_j=\sigma\}\cap\{u_i>\delta\}}\frac{\partial u_i}{\partial v_{i,2}^-} \\
&\quad +\sum_{i\neq j}\int_{\{u_j=\delta\}\cap\{v_i>\sigma\}}|\nabla u_j(x)|v_i^\sigma \\
&\quad -2\sum_{i,j}\delta\int_{\{u_i=\delta\}\cap\{v_j<\sigma\}}\frac{\partial u_i}{\partial v_{i,1}^-} - 2\sum_{i,j}\int_{\{u_i>\delta\}\cap\{v_j=\sigma\}}\frac{\partial u_i}{\partial v_{i,2}^+}u_i. \tag{3.17}
\end{aligned}
$$

In these twelve terms, let us first look at the second, the seventh and the twelfth terms (modulo the constant -2):

$$\sum_i \int_{\{v_i=\sigma\}\cap\{u_i>\delta\}} \frac{\partial u_i}{\partial v_{i,2}^-} u_i + \sum_{i\neq j} \int_{\{v_j=\sigma\}\cap\{u_i>\delta\}} \frac{\partial u_i}{\partial v_{i,2}^-} u_i$$

$$+ \sum_{i,j} \int_{\{u_i>\delta\}\cap\{v_j=\sigma\}} \frac{\partial u_i}{\partial v_{i,2}^+} u_i$$

$$= \sum_i \sum_j \int_{\{v_i=\sigma\}\cap\{u_j>\delta\}} \frac{\partial u_i}{\partial v_{i,2}^-} u_i + \sum_{i,j} \int_{\{v_i=\sigma\}\cap\{u_j>\delta\}} \frac{\partial u_i}{\partial v_{i,2}^+} u_i$$

$$= 0,$$

because the normal vector field $v_{i,2}^+$ and $v_{i,2}^-$ on $\{v_i=\sigma\}$ have opposite directions.

The fourth and the ninth terms are

$$2 \sum_i \sigma \int_{\{v_i=\sigma\}\cap\{u_i>\delta\}} \frac{\partial u_i}{\partial v_{i,2}^-}$$

and

$$-2 \sum_{i\neq j} \sigma \int_{\{v_j=\sigma\}\cap\{u_i>\delta\}} \frac{\partial u_i}{\partial v_{i,2}^-}.$$

We show that the integration in these terms are uniformly bounded in σ, thus as $\sigma \to 0$, these two terms converge to 0. We only calculate the first one and the second one is similar. First,

$$\int_{\{v_i=\sigma\}\cap\{u_i>\delta\}} \frac{\partial u_i}{\partial v_{i,2}^-} = \int_{\{v_i>\sigma\}\cap\{u_i>\delta\}} \Delta u_i - \int_{\{v_i>\sigma\}\cap\{u_i=\delta\}} \frac{\partial u_i}{\partial v_{i,1}^-}$$

$$- \int_{\{v_i>\sigma\}\cap\{u_i>\delta\}\cap\partial\Omega'} \frac{\partial u_i}{\partial v},$$

where v is the outward unit normal vector field to $\partial\Omega'$. In the right-hand side, the first term is less than the total mass of the measure Δu_i on Ω; the second term can be controlled by

$$\int_{\{u_i=\delta\}} |\nabla u_i|,$$

which will be shown to be uniformly bounded in δ at the end of this section; the third term is also uniformly bounded by the area of $\partial\Omega'$ times the sup norm of ∇u_i, and we conclude.

The eleventh term in Eq. (3.17) converges to 0 as $\sigma \to 0$. Now we can take the limit in the remaining terms as $\sigma \to 0$ to get:

$$
2\sum_i \int_{\{u_i=\delta\}\cap\{v_i>0\}} \frac{\partial u_i}{\partial v_{i,1}^-} v_i - 2\delta \sum_i \int_{\{u_i=\delta\}\cap\{v_i>0\}} \frac{\partial u_i}{\partial v_{i,1}^-}
$$

$$
- \sum_{i\neq j} \int_{\{u_i=\delta\}\cap\{v_j>0\}} \big|\nabla u_i(x)\big| v_j - 2\sum_{i\neq j} \int_{\{u_i=\delta\}\cap\{v_j>0\}} \frac{\partial u_i}{\partial v_{i,1}^-} v_j
$$

$$
- 2\delta \sum_{i\neq j} \int_{\{u_i=\delta\}\cap\{v_j>0\}} \frac{\partial u_i}{\partial v_{i,1}^-} + \sum_{i\neq j} \int_{\{u_j=\delta\}\cap\{v_i>0\}} \big|\nabla u_j(x)\big| v_i .
$$

Noting that on $\{u_i=\delta\}$, $\frac{\partial u_i}{\partial v_{i,1}^-} = -|\nabla u_i|$, so we get

$$
-2\sum_i \int_{\{u_i=\delta\}\cap\{v_i>0\}} \big|\nabla u_i\big| v_i - \sum_{i\neq j} \int_{\{u_i=\delta\}\cap\{v_j>0\}} \big|\nabla u_i(x)\big| v_j
$$

$$
+2\sum_{i\neq j} \int_{\{u_i=\delta\}\cap\{v_j>0\}} \big|\nabla u_i\big| v_j + \sum_{i\neq j} \int_{\{u_j=\delta\}\cap\{v_i>0\}} \big|\nabla u_j(x)\big| v_i
$$

$$
+2\delta \sum_{i\neq j} \int_{\{u_i=\delta\}\cap\{v_j>0\}} \big|\nabla u_i\big| + 2\delta \sum_i \int_{\{u_i=\delta\}\cap\{v_i>0\}} \big|\nabla u_i\big|.
$$

The integration in the last two terms will be shown to be uniformly bounded in δ at the end of this subsection, thus as $\delta \to 0$, they converge to 0.

As $\delta \to 0$, the remaining terms converge to (see Sect. 3.2.1)

$$
-2\sum_i \int_{\partial\{u_i>0\}\cap\{v_i>0\}} \big|\nabla u_i\big| v_i - \sum_{i\neq j} \int_{\partial\{u_i>0\}\cap\{v_j>0\}} \big|\nabla u_i(x)\big| v_j
$$

$$
+2\sum_{i\neq j} \int_{\partial\{u_i>0\}\cap\{v_j>0\}} \big|\nabla u_i\big| v_j + \sum_{i\neq j} \int_{\partial\{u_j>0\}\cap\{v_i>0\}} \big|\nabla u_j(x)\big| v_i .
$$

In $\{v_i>0\}$, if $\partial\{u_i>0\}\cap\partial\{u_j>0\}\neq\emptyset$, then $\frac{\partial u_i}{\partial v_{i,1}} = \frac{\partial u_j}{\partial v_{j,1}}$, so the first term cancels some terms in the third term, with

$$
\sum_i \sum_{j,k\neq i} \int_{\partial\{u_j>0\}\cap\partial\{u_k>0\}\cap\{v_i>0\}} \big|\nabla u_j(x)\big| v_i
$$

left. The integral $\int_{\partial\{u_j>0\}\cap\{v_i>0\}} |\nabla u_j(x)| v_i$ appears twice in the second term and the fourth term with opposite signs, so these terms cancel each other.

So we have

$$
\left. \frac{dE}{dt} \right|_{t=0} \geq 0.
$$

However, $E(t)$ is a convex function of t, so 0 is its minimizer. But from our choice of v, we also have 1 as its minimizer. Therefore, we must have $E(t) \equiv const.$, this implies that u is the energy minimizer and $u \equiv v$.

3.2.1 Verification of the Convergence of the Integration

Here, we will show the uniform boundedness and the convergence of various integrations appearing before.

We only consider the integration

$$\int_{\{u_i=\delta\}\cap\{v_i>0\}} |\nabla u_i| v_i,$$

and others can be treated similarly.

We know that the singular set of $\partial\{u_i > 0\}$ is of Hausdorff dimension $n-2$, so its $n-1$ dimensional Hausdorff measure is 0. In particular, $\forall \varepsilon > 0$, there exists some balls $B(x_k, r_k)$, such that

$$Sing\big(\partial\{u_i > 0\}\big) \subset \bigcup B(x_k, r_k),$$

with

$$\sum_k r_k^{n-1} \le \varepsilon.$$

Outside $\bigcup B(x_k, r_k)$, $\partial\{u_i > 0\}$ is a regular smooth hypersurface where $\inf |\nabla u_i| > 0$, then it is easily seen that (noting that the integrand are continuous up to $\partial\{u_i > 0\}$ outside $\bigcup B(x_k, r_k)$), there exists a $\delta_0 > 0$, such that for any $\delta < \delta_0$ (if the level surface $\{u_i = \delta\}$ is regular)

$$\left| \int_{\{u_i=\delta\}\setminus(\bigcup B(x_k,r_k))} |\nabla u_i| v_i(x) - \int_{\partial\{u_i>0\}\setminus(\bigcup B(x_k,r_k))} |\nabla u_i| v_i(x) \right| \le \varepsilon.$$

On the other hand, in $B(x_k, r_k)$, we have

$$0 = \int_{\{u_i>\delta\}\cap B(x_k,r_k)} \Delta u_i$$

$$= -\int_{\{u_i=\delta\}\cap B(x_k,r_k)} |\nabla u_i(x)| + \int_{\{u_i>\delta\}\cap \partial B(x_k,r_k)} \frac{\partial u_i}{\partial \nu},$$

where ν is the unit outward normal vector of $\partial B(x_k, r_k)$. We also have on $\{u_i > \delta\} \cap \partial B(x_k, r_k)$

$$\left| \frac{\partial u_i}{\partial \nu} \right| \le |\nabla u_i| \le C.$$

Combing these two facts, we get

$$\int_{\{u_i=\delta\}\cap B(x_k,r_k)} |\nabla u_i(x)| \le C r_k^{n-1}.$$

Sum these to get

$$\sum_k \int_{\{u_i=\delta\}\cap B(x_k,r_k)} |\nabla u_i(x)| \le C\varepsilon.$$

From $H^{n-1}(\partial\{u_i > 0\} \cap \Omega') < \infty$, we can select balls $B(x_k, r_k)$ small enough so that (here the integration, as usual, is understood as on the regular part)

$$\sum_k \int_{\partial\{u_i>0\}\cap B(x_k,r_k)} |\nabla u_i(x)| \le \varepsilon.$$

In view of the arbitrary choice of ε, now it is clear that as $\delta \to 0$,

$$\int_{\{u_i=\delta\}\cap\{v_i>0\}} |\nabla u_i| v_i(x) \to \int_{\partial\{u_i>0\}\cap\{v_i>0\}} |\nabla u_i| v_i(x).$$

Then the left-hand side is also uniformly bounded in δ.

3.3 Uniqueness and Asymptotics for the Singular Parabolic System

Finally, we consider the corresponding parabolic problem of (3.1).

$$\begin{cases} \dfrac{\partial u_i}{\partial t} - \Delta u_i \le 0, & \text{in } \Omega \times (0,+\infty), \\[2mm] \left(\dfrac{\partial}{\partial t} - \Delta\right)\left(u_i - \sum_{j\neq i} u_j\right) \ge 0, & \text{in } \Omega \times (0,+\infty), \\[2mm] u_i u_j = 0, & \text{in } \Omega \times (0,+\infty), \\[2mm] u_i = \varphi_i, & \text{on } \partial\Omega \times (0,+\infty), \\[2mm] u_i = \phi_i, & \text{on } \Omega \times \{0\}. \end{cases} \qquad (3.18)$$

As before, the solution is understood in the H^1 sense. We have the following uniqueness result.

Theorem 3.3.1 *There exists a unique solution for the problem* (3.18). *Moreover, as* $t \to +\infty$, *this solution converges to the unique solution of Eq.* (3.1).

First, we prove the uniqueness of the solution. If there exist two solutions of (3.18), u and v, define the distance

$$d(u, v)(x, t) := \sum_i \left| u_i(x, t) - v_i(x, t) \right|.$$

Because $\forall i$, u_i and v_i are Lipschitz continuous with respect to the parabolic distance, d is Lipschitz too. Now we claim that in $\{d > 0\}$

$$\left(\Delta - \frac{\partial}{\partial t} \right) d \geq 0.$$

We prove this case by case:

1. Where $u_i > 0$ and $v_i > 0$ with $u_i - v_i > 0$, we have

$$d = u_i - v_i,$$

so $(\Delta - \frac{\partial}{\partial t})d = 0$.
2. Where $u_i > 0$ and $v_i > 0$ with $v_i - u_i > 0$, we have

$$d = v_i - u_i,$$

so $(\Delta - \frac{\partial}{\partial t})d = 0$.
3. Where $u_i > 0$ and $v_j > 0$ for some $j \neq i$, we have

$$d = v_j + u_i,$$

so $(\Delta - \frac{\partial}{\partial t})d = 0$.
4. Where $u_i(X_0) > 0$ and $v_j(X_0) = 0$, $\forall j$, then in a neighborhood of X_0, we have

$$d = u_i - v_i + \sum_{j \neq i} v_j,$$

so

$$\left(\Delta - \frac{\partial}{\partial t} \right) d = \left(\Delta - \frac{\partial}{\partial t} \right) \left(-v_i + \sum_{j \neq i} v_j \right) \geq 0.$$

5. Where $v_i(X_0) > 0$ and $u_j(X_0) = 0$, $\forall j$, then in a neighborhood of X_0, we have

$$d = v_i - u_i + \sum_{j \neq i} u_j,$$

so

$$\left(\Delta - \frac{\partial}{\partial t} \right) d = \left(\Delta - \frac{\partial}{\partial t} \right) \left(-u_i + \sum_{j \neq i} u_j \right) \geq 0.$$

Take a $\varepsilon > 0$, and define

$$\hat{d} = e^{-\varepsilon t} d,$$

then

$$\left(\Delta - \frac{\partial}{\partial t}\right)\hat{d} > 0,$$

strictly on the open set $\{d > 0\}$. Now from the boundary condition we have

$$\hat{d} = d = 0, \quad \text{on } \partial_p\left(\Omega \times (0, +\infty)\right).$$

By the maximum principle we get $\hat{d} \equiv 0$, or in other words

$$u_i \equiv v_i.$$

That is, the solution is unique.

Next, let us consider the singular limit of the following system (this was considered by Caffarelli and Lin in [5]):

$$\begin{cases} \dfrac{\partial v_{i,\kappa}}{\partial t} - \Delta v_{i,\kappa} = -\kappa v_{i,\kappa} \displaystyle\sum_{j \neq i} v_{j,\kappa}^2, & \text{in } \Omega \times (0, +\infty), \\[2mm] v_{i,\kappa} = \varphi_i, & \text{on } \partial\Omega \times (0, +\infty), \\[2mm] v_{i,\kappa} = \phi_i, & \text{on } \Omega \times \{0\}. \end{cases} \tag{3.19}$$

This system is the descent gradient flow of the functional

$$\int_\Omega 2\sum_i |\nabla v_i|^2 + \kappa \sum_{i \neq j} v_i^2 v_j^2.$$

We claim that its singular limit v_i as $\kappa \to +\infty$ satisfy the inequalities in (3.18). We know that the singular limit satisfy

$$\frac{\partial v_i}{\partial t} - \Delta v_i = -\sum_{j \neq i} \mu_{ij},$$

where μ_{ij} are positive Radon measure supported on $\partial\{v_i > 0\} \cap \partial\{v_j > 0\}$. We just need to show

$$\mu_{ij} = \mu_{ji}, \quad \forall j \neq i.$$

This comes from the regularity theory of the free boundary, which shows that

$$\mu_{ij} = |\nabla v_i| H^{n-1}\big\lfloor_{\partial\{v_i > 0\}} \wedge dt.$$

But on $\partial\{v_i > 0\} \cap \partial\{v_j > 0\}$, we have

$$|\nabla v_i| = |\nabla v_j|, \quad H^{n-1} \quad \text{a.e.}$$

Then our claim is proven. For more details, see Chap. 7.

From the above proof of the uniqueness, we know this singular limit v coincides with u, the solution of (3.18). But v has an energy inequality induced from (3.19):

$$\frac{d}{dt} \int_\Omega \sum_i |\nabla v_i|^2 \leq - \int_\Omega \sum_i \left|\frac{\partial v_i}{\partial t}\right|^2.$$

Of course, this is also valid for u. Now it is easy to conclude that as $t \to +\infty$, u converge to the unique stationary solution. This is because, for any sequence $t_i \to +\infty$, the translation $u(t_i + t)$ has a subsequence converges to a solution w of (3.18) defined on $(-\infty, +\infty)$. However, from the energy decreasing property, we know

$$0 \geq -\frac{d}{dt} \int_\Omega \sum_i |\nabla w_i|^2 \geq \int_\Omega \sum_i \left|\frac{\partial w_i}{\partial t}\right|^2.$$

So

$$\frac{\partial w_i}{\partial t} = 0, \quad \text{a.e.}$$

that is, w is a stationary solution of (3.18), or solution of (3.1). From Theorem 3.2.1, we know such w is unique, thus we proved that for any sequence $t_i \to +\infty$

$$u(t_i) \to w,$$

with w the unique solution of (3.1).

Remark 3.3.2 The above method can be easily generalized to a more general setting. In particular, there exists a unique solution to the following initial-boundary value problem

$$
\begin{cases}
\dfrac{\partial u_i}{\partial t} - \Delta u_i \leq f_i(u_i), & \text{in } \Omega \times (0, +\infty), \\[2mm]
\left(\dfrac{\partial}{\partial t} - \Delta\right)\left(u_i - \sum_{j \neq i} u_j\right) \geq f_i(u_i) - \sum_{j \neq i} f_j(u_j), & \text{in } \Omega \times (0, +\infty), \\[2mm]
u_i u_j = 0, & \text{in } \Omega \times (0, +\infty), \\[2mm]
u_i = \varphi_i, & \text{on } \partial\Omega \times (0, +\infty), \\[2mm]
u_i = \phi_i, & \text{on } \Omega \times \{0\},
\end{cases}
$$

where $f_i(u_i)$ are Lipschitz continuous functions on $\mathbb{R}$.

Chapter 4
The Dynamics of One Dimensional Singular Limiting Problem

Abstract In this chapter we study the dynamics of solutions to the singular parabolic system (1.12). We prove that every solution will converge to a stationary solution of (1.12) as time goes to infinity. We also give some interesting properties of these solutions, see Sect. 4.2. The stationary problem is also studied in Sect. 4.1.

4.1 The Stationary Case

In this section, we study the stationary case of differential inequalities (1.12):

$$
\begin{cases}
-\dfrac{d^2 u_i}{dx^2} \leq a_i u_i - u_i^2 & \text{in } (0,1); \\[2mm]
-\dfrac{d^2}{dx^2}\left(u_i - \sum_{j \neq i} u_j\right) \geq a_i u_i - u_i^2 - \sum_{j \neq i}(a_j u_j - u_j^2) & \text{in } (0,1); \\[2mm]
u_i \geq 0, \qquad u_i(0) = u_i(1) = 0; \\[2mm]
u_i u_j = 0 \quad \text{in } [0,1], \text{ for } i \neq j \ (1 \leq i, j \leq M).
\end{cases}
\tag{4.1}
$$

The main result of this section is Theorem 4.1.4. The Neumann boundary value problem can be treated similarly. We know that the above problem can have solutions with some identically zero components, thus it may have multiple solutions, but if the set where u_i is positive has several components, we treat these as different species (because we are in the stationary case). Here, the number of species may become infinite. (Below we will exclude this possibility.) We have the following theorem. (Note that Lemma 4.1.2 below ensures that the set where u_i is positive has only finitely many components.)

Theorem 4.1.1 $\forall M \in \mathbb{N}$, there exists at most one solution $(u_1, u_2, \ldots, u_M)$ of (4.1) (up to permutation) such that each u_i is not identically zero and each of its supports is an interval.

First, we need a well-known lemma. For later use and completeness, we give the proof here.

K. Wang, *Free Boundary Problems and Asymptotic Behavior of Singularly
Perturbed Partial Differential Equations*, Springer Theses,
DOI 10.1007/978-3-642-33696-6_4, © Springer-Verlag Berlin Heidelberg 2013

Lemma 4.1.2 *For each $L > \frac{\pi}{2\sqrt{a}}$, there exists a unique positive solution u of*

$$
\begin{cases}
-\dfrac{d^2u}{dx^2} = au - u^2, & in\ (0, L) \\
u(0) = u(L) = 0.
\end{cases}
\tag{4.2}
$$

There is no positive solution if $L \leq \frac{\pi}{2\sqrt{a}}$.

Proof The existence can be easily proven by, for example, the method of sub- and sup-solutions. In fact, we have the following conservation quantity:

$$
\left(\frac{du}{dx}\right)^2 + au^2 - \frac{2}{3}u^3 \equiv c,
\tag{4.3}
$$

for some constant c.

If we have two positive solutions u and v, then on the open set $D := \{u > v\}$ (if not empty), we have

$$
\begin{aligned}
0 \geq \int_{\partial D} \frac{\partial u}{\partial v}v - \frac{\partial v}{\partial v}u &= \int_D \Delta uv - \Delta vu \\
&= \int_D -(au - u^2)v + (av - v^2)u \\
&= \int_D uv(u - v) \\
&> 0,
\end{aligned}
$$

which is a contradiction. $\qquad\qquad\square$

Corollary 4.1.3 *The constant c defined in (4.3), seen as a function of L, is increasing in L.*

Proof Assume $L_1 > L_2 > \frac{\pi}{2\sqrt{a}}$ with u_1 and u_2 the solution of (4.2) in $[0, L_1]$ and $[0, L_2]$, respectively. Because

$$
0 = u_2(L_2) < u_1(L_2),
$$

the same method as in the previous lemma gives that in $[0, L_2]$

$$
u_1 > u_2.
$$

This implies

$$
\frac{du_1}{dx}(0) > \frac{du_2}{dx}(0).
$$

Note that here we have the strict inequality because otherwise we will have $u_1 \equiv u_2$. Because

$$c_i(L_i) = \left(\frac{du_i}{dx}(0)\right)^2,$$

our claim follows. $\qquad\qquad\qquad\qquad\qquad\qquad\qquad\qquad\qquad\qquad\qquad$ □

Now we give the proof of Theorem 4.1.1.

Proof Since each support $\{u_i > 0\}$ is an interval, and for $i \neq j$, $\{u_i > 0\}$ and $\{u_j > 0\}$ are disjoint (recall that $u_i u_j \equiv 0$), we can assume

$$\{u_i > 0\} = (\alpha_i, \beta_i),$$

with (by renumbering if necessary)

$$0 \leq \alpha_1 < \beta_1 \leq \alpha_2 < \beta_2 \leq \cdots \leq \alpha_m < \beta_m \leq 1.$$

Note that on the set $(\beta_{i-1}, \alpha_{i+1})$, where all $u_j \equiv 0$ for $j \neq i$, u_i satisfies the equation (by combining the first and second equation of (4.1))

$$-\frac{d^2 u_i}{dx^2} = a_i u_i - u_i^2. \tag{4.4}$$

Moreover, $u_i \geq 0$ and u_i does not vanish identically on this interval, so we must have $u_i > 0$ in $(\beta_{i-1}, \alpha_{i+1})$. Thus

$$\alpha_{i+1} = \beta_i, \quad \text{and} \quad \alpha_1 = 0, \ \beta_m = 1,$$

that is, the set where all u_i vanish contains at most finitely many points.

In (α_i, β_i), there exists a constant $c_i > 0$ such that

$$\left(\frac{du_i}{dx}\right)^2 + a_i u_i^2 - \frac{2}{3} u_i^3 = c_i. \tag{4.5}$$

Because $u_i(\alpha_i) = u_i(\beta_i) = 0$, so

$$\left(\frac{du_i}{dx}(\alpha_i)\right)^2 = c_i = \left(\frac{du_i}{dx}(\beta_i)\right)^2.$$

In (α_i, β_{i+1}), from the second inequalities in (4.1)

$$\begin{cases} -\dfrac{d^2}{dx^2}(u_i - u_{i+1}) \geq a_i u_i - u_i^2 - \left(a_{i+1} u_{i+1} - u_{i+1}^2\right) & \text{in } (\alpha_i, \beta_{i+1}); \\[2mm] -\dfrac{d^2}{dx^2}(u_{i+1} - u_i) \geq a_{i+1} u_{i+1} - u_{i+1}^2 - \left(a_i u_i - u_i^2\right) & \text{in } (\alpha_i, \beta_{i+1}); \end{cases}$$

we see that

$$-\frac{d^2(u_i - u_{i+1})}{dx^2} = a_i u_i - u_i^2 - a_{i+1}u_{i+1} + u_{i+1}^2.$$

This implies $u_i - u_{i+1} \in C^1(\alpha_i, \beta_{i+1})$. Then we have

$$\left(\frac{du_i}{dx}(\beta_i)\right)^2 = \left(\frac{du_{i+1}}{dx}(\beta_i)\right)^2.$$

So all of the c_i in (4.5) are the same, which we denote by c. Now from Corollary 4.1.3, we know that each $\beta_i - \alpha_i$ is uniquely determined by c (because it is a strictly increasing function of c), then there exists at most one c such that

$$\sum_i (\beta_i - \alpha_i) = 1.$$

□

Theorem 4.1.4 *There are only finitely many solutions of* (4.1).

Proof It is well known that if $u > 0$ in a bounded domain Ω with boundary value 0, and

$$-\Delta u = au - u^2,$$

then

$$a > \lambda_1,$$

where λ_1 is the first eigenvalue of Ω with Dirichlet boundary condition.

Now in our case, each connected component of u_i, which is an interval, has length $\geq \frac{\pi}{2\sqrt{a_i}}$. So the total number of connected components of each possible solution of (4.1) has an upper bound, which depends on those a_i, $1 \leq i \leq M$. By Theorem 4.1.1, we can finish the proof.   □

4.2 The Parabolic Case

In this section, we study the dynamics of the following one dimensional singular parabolic problem.

$$\begin{cases} \dfrac{\partial u_i}{\partial t} - \dfrac{\partial^2 u_i}{\partial x^2} \leq a_i u_i - u_i^2 & \text{in } (0,1) \times (0, +\infty); \\[2mm] \left(\dfrac{\partial}{\partial t} - \dfrac{\partial^2}{\partial x^2}\right)\left(u_i - \displaystyle\sum_{j \neq i} u_j\right) \geq a_i u_i - u_i^2 - \displaystyle\sum_{j \neq i}\left(a_j u_j - u_j^2\right) \\[2mm] \quad \text{in } (0,1) \times (0, +\infty); \\[2mm] u_i = \phi_i, \quad \text{on } [0,1] \times \{0\}; \\[2mm] u_i u_j = 0, \quad \text{on } [0,1] \times (0, +\infty). \end{cases} \qquad (4.6)$$

Here $i = 1, 2, \ldots, M$, and ϕ_i are given nonnegative Lipschitz continuous functions on $[0, 1]$, such that

$$\phi_i \geq 0, \quad \text{and if} \quad i \neq j, \ \phi_i \phi_j \equiv 0.$$

For simplicity, we take those b_{ij} in (1.1) to be 1. It will be clear from the proof that the general case can be treated similarly. The main result of this section is summarized in the following theorem.

Theorem 4.2.1 *Let*

$$m(T) := \text{The number of connected components of } \bigcup_i \{u_i > 0\} \cap \{t = T\},$$

then

1. *$m(T)$ is non-increasing in T;*
2. *there exist at most countable $T_1 > T_2 > \cdots$, satisfying $\lim_{k \to \infty} T_k = 0$, such that these points are exactly those times where the value of $m(T)$ jumps;*
3. *for $t \in (T_{i+1}, T_i)$ (here we take the convention that $T_0 = +\infty$), after re-indexing i and ignoring those u_i which are identically 0 in this time interval, we can define*

$$u := u_1 - u_2 + u_3 - u_4 + \cdots .$$

   u satisfies

$$\left(\frac{\partial}{\partial t} - \frac{\partial^2}{\partial x^2} \right) u = f. \tag{4.7}$$

   Here $f := a_1 u_1 - u_1^2 - (a_2 u_2 - u_2^2) + \cdots$ is a Lipschitz continuous function.
4. *In $[0, 1] \times (T_1, +\infty)$, we have $\lim_{t \to +\infty} u_i(x, t) = v_i(x)$, where $v := (v_1, v_2, \ldots)$ is a stationary solution of (4.6). Moreover, either $v \equiv 0$ or v has exactly $m(T_1)$ nonzero components.*

With similar ideas, we also give a description of connecting orbits (Proposition 4.2.12) and a uniqueness result of the initial value problem (Proposition 4.2.14).

We first show a regularizing property of (4.6). This is a special case of the Dimension Reduction Principle (see for example [7]). However, for this problem it is not written down explicitly in the literature, so we will give the arguments here for completeness.

Proposition 4.2.2 *For each $t > 0$, for the free boundary $\mathscr{F}(u) := \bigcup_i \partial \{u_i > 0\}$ in the problem of (4.6), $\mathscr{F}(u) \cap \{t\}$ contains only finitely many points.*

Proof Assume for some $t > 0$, there exist infinitely many points in $\mathscr{F}(u) \cap \{t\}$. Take a $x_0 \in \mathscr{F}(u) \cap \{t\} \cap (0, 1)$, such that there exist $x_k \neq x_0$ in $\mathscr{F}(u)$ and $\lim_{k \to \infty} x_k = x_0$. Without loss of generality, we can assume $x_0 = 0$ and $t = 0$. (We can change the domain by translation.)

Take

$$r_k = x_k - x_0,$$

which we can assume to be a positive number. Now define the blow up sequence

$$u^k(x, t) := \frac{1}{L_k} u(r_k x, r_k^2 t),$$

where $L_k = \frac{1}{|C_{r_k}|} \int_{C_{r_k}} |u|^2$ and $C_{r_k} = [-r_k, r_k] \times [-r_k^2, r_k^2]$ is the parabolic cylinder.

As in [2] (or Theorem 2.1 in [7]), we know there exists a subsequence (still denoted by u^k), converging to $v(x, t)$ locally uniformly on $\mathbb{R}^1 \times (-\infty, +\infty)$, with v nontrivial, nonnegative. Moreover, there exists a positive integer d, which is the frequency of $(0, 0)$ (for the definition and its property, please see Sect. 9 in [2]), such that in $\mathbb{R}^1 \times (-\infty, +\infty)$

$$
\begin{cases}
\dfrac{\partial v_i}{\partial t} - \dfrac{\partial^2 v_i}{\partial x^2} \leq 0, \\[2mm]
\left(\dfrac{\partial}{\partial t} - \dfrac{\partial^2}{\partial x^2} \right) \left(v_i - \displaystyle\sum_{j \neq i} v_j \right) \geq 0, \\[2mm]
v_i v_j = 0 \quad \text{for } i \neq j, \\[2mm]
v_i(\lambda x, \lambda^2 t) = \lambda^d v_i(x, t).
\end{cases}
\tag{4.8}
$$

Because $(1, 0) \in \mathscr{F}(u^k)$, we also have $v_i(1, 0) = 0$ for all i, or equivalently $(1, 0) \in \mathscr{F}(v)$. Then from the homogeneity of v, we have

$$\mathbb{R}_+^1 \times \{0\} \subset \mathscr{F}(v).$$

This contradicts the dimension estimate of the free boundary (from Sect. 10 of [2], and Theorem 2.3 in [7], we know the Hausdorff dimension of the free boundary restricted to time $t = 0$ is 0).

If x_0 lies on the fixed boundary, the blow up limit v is defined on $\mathbb{R}_+^1 \times (-\infty, +\infty)$, with the boundary condition

$$v = 0, \quad \text{on } \{0\} \times (-\infty, +\infty).$$

This can be treated similarly.   □

Now we have the following lemma.

Lemma 4.2.3 $\forall T_1 > T_2 > 0$, *any connected component of* $\{u_i > 0\} \cap \{t < T_1\}$ *must intersect with* $\{t = T_2\}$.

Proof We know that $\{u_i > 0\}$ is an open subset of $[0, 1] \times [0, +\infty)$, so it is locally path connected, and the path connected components coincide with the connected components. Assuming there exists a connected component A, with $\{t = T_2\} \cap A = \emptyset$.

Now consider the restriction $\hat{u}_i$ of u_i on A. First, it is still continuous. So we have

$$\begin{cases} \dfrac{\partial \hat{u}_i}{\partial t} - \dfrac{\partial^2 \hat{u}_i}{\partial x^2} \leq a_i \hat{u}_i - \hat{u}_i^2, & \text{in } \{\hat{u}_i > 0\}, \\ \hat{u}_i \geq 0, & \text{in } [0, 1] \times [T_2, T_1], \\ \hat{u}_i = 0, & \text{on } \partial_p\big([0, 1] \times [T_2, T_1]\big), \end{cases} \tag{4.9}$$

where $\partial_p([0, 1] \times [T_2, T_1])$ is the parabolic boundary. Now, we must have $\hat{u}_i \equiv 0$. In fact, take a positive constant $C > a_i$, and define

$$v(x, t) := e^{-Ct}\hat{u}_i(x, t).$$

We have in the open set $\{v > 0\}$

$$\frac{\partial v}{\partial t} - \frac{\partial^2 v}{\partial x^2} \leq e^{-Ct}\big(-C\hat{u}_i + a_i\hat{u}_i - \hat{u}_i^2\big) < 0.$$

If $\hat{u}_i$ is not identically zero, then $\max_{[0,1]\times[T_2,T_1]} v$ is attained at some point (x_0, t_0) where $\hat{u}_i(x_0, t_0) > 0$. We know $\hat{u}_i$ is positive in a neighborhood of (x_0, t_0) because $\hat{u}_i$ is continuous. Then v is smooth in this neighborhood and we can apply the maximum principle to get a contradiction. $\qquad \Box$

Remark 4.2.4 From the proof, we see this lemma remains true in higher dimensions and with a more general nonlinearity.

Without loss of generality, we can assume at $t = T' > 0$, u_i satisfy the condition: $\{u_i > 0\} \cap \{t = T'\}$ is an interval for each i. Later it will be clear that, as time evolves, two connected components may merge into one, but a single connected component cannot split into two. With this assumption in hand, we have the following corollary.

Corollary 4.2.5 $\forall T > T'$, $\{u_i > 0\} \cap \{t = T\}$ *is an interval if it is not empty.*

Proof Assume $\exists T > 0$, such that $\{u_i > 0\} \cap \{t = T\}$ contains two disjoint intervals $(\alpha_1, \beta_1) \cup (\alpha_2, \beta_2)$ where

$$\alpha_1 < \beta_1 < \alpha_2 < \beta_2,$$

but

$$(\beta_1, \alpha_2) \cap \{u_i > 0\} = \emptyset.$$

From Lemma 4.2.3, we know there exist two continuous paths $\gamma_1(t)$ and $\gamma_2(t)$ which start from a point in $\{u_i > 0\} \cap \{t = T'\}$ and end in two points lying in $(\alpha_1, \beta_1) \times \{T\}$ and $(\alpha_2, \beta_2) \times \{T\}$ respectively.

There exists some $j \neq i$ such that

$$\big((\beta_1, \alpha_2) \times \{T\}\big) \cap \{u_j > 0\} \neq \emptyset.$$

However, this set is contained in a connected component which can not be connected to $\{t = T'\}$ through a continuous path. This contradicts Lemma 4.2.3 and we can conclude.   □

Now the first and the second part of Theorem 4.2.1 can be easily seen. The points T_k are exactly where some components of u_i become extinct.

Assume in (T_{i+1}, T_i), $u_1, u_2, \ldots, u_m$ are not identically zero, where $m \leq M$. With the help of Lemma 4.2.3, we can show that, any two connected components of $\{u_i > 0\}$ and $\{u_j > 0\}$ with $i \neq j$ cannot interweave, that is, if at $t_1 \in (T_{i+1}, T_i)$, $\{u_i > 0\}$ lies above $\{u_j > 0\}$, at another $t_2 \in (T_{i+1}, T_i)$, $\{u_i > 0\}$ can not lie below $\{u_j > 0\}$. So we can rearrange the index properly so that $\{u_i > 0\}$ is adjacent to $\{u_{i-1} > 0\}$ and $\{u_{i+1} > 0\}$ (here we treat the same species with different domains as different species, this causes no confusions in (T_{i+1}, T_i)). This proves the third part of Theorem 4.2.1. Now in this time interval we have the following lemma.

Lemma 4.2.6 $\forall t \in (T_{i+1}, T_i)$, *there exist*

$$0 = \alpha_0(t) \leq \alpha_1(t) \leq \cdots < \alpha_m(t) = 1,$$

such that

$$\{u_j > 0\} \cap \{t\} = (\alpha_{j-1}(t), \alpha_j(t)).$$

Moreover, all the $\alpha_j(t)$ are continuous functions of t.

Proof From Corollary 4.2.5, we know that for each $t > 0$, there exist

$$0 = \alpha_0(t) < \beta_1(t) \leq \alpha_1(t) < \beta_2(t) \leq \cdots < \beta_m(t) = 1,$$

such that

$$\{u_j > 0\} \cap \{t\} = (\alpha_{j-1}(t), \beta_j(t)).$$

From the unique continuation property of the linear parabolic equation

$$\frac{\partial u}{\partial t} - \frac{\partial^2 u}{\partial x^2} = Vu, \qquad (4.10)$$

where

$$V := (a_1 - u_1) - (a_2 - u_2) + \cdots$$

is a L^∞ function, we know that, $\forall t$, the set $\{x : u(x, t) = 0\}$ cannot contain an open set. From this, we easily see

$$\beta_j(t) = \alpha_j(t).$$

Now $\bigcup_i \{(\alpha_j(t), t) : t \in (T_{i+1}, T_i)\}$ is the nodal set of u, and the continuity of $\alpha_j(t)$ is easily seen by its local uniqueness.   □

Remark 4.2.7 In the proof, we need the fact that u is not identically zero in $[0, 1] \times (T_{i+1}, T_i)$. This can be guaranteed by the unique continuation property of the linear parabolic equation (4.9) (see [38], note here we have zero boundary condition). Note that, although the form of equation changes when crossing those extinction time T_i, if $u_i \equiv 0$ in $\{t > T_i\}$, then $u_i \equiv 0$ at $t = T_i$ and we can apply the backward uniqueness in $(T_{i+1}, T_i]$ again.

These curves are in fact regular, that is the following proposition.

Proposition 4.2.8 $\forall t \in (T_{i+1}, T_i)$, *at* $(\alpha_j(t), t)$, *we have (here u defined as in Theorem* 4.2.1)

$$\frac{\partial u}{\partial x} \neq 0.$$

In particular, $\alpha_j(t)$ are C^1 in t.

Proof Note that since u satisfies (4.10), and both V and u are bounded, standard parabolic estimates imply that u is C^1 in the space variable x. So the formulation in the proposition makes sense.

We need the characterization of the singular points of free boundaries (see Sect. 9 of [2] or [7]). Assuming we are at $(0, 0)$, we have a positive integer number d (the frequency) such that the blow up sequences

$$u_\lambda(x, t) = \lambda^{-d} u(\lambda^2 t, \lambda x)$$

converge to $v(x, t)$ as $\lambda \to 0$, which is a polynomial of the form (Hermitian polynomial)

$$\sum_{k=0}^{[\frac{d}{2}]} \frac{d!}{k!(d - 2k)!} x^{d-2k} t^k.$$

Moreover, if $\frac{\partial u}{\partial x} = 0$, then $d \geq 2$. In other words, $\frac{\partial u}{\partial x} \neq 0$ if and only if $d = 1$.

Now, $\{v(x, t) = 0\}$ is composed of d nonsingular curves C_i which have the form

$$t = -c_i x^2, \quad 1 \leq i \leq d.$$

Here c_i are the zeros of the corresponding Hermite polynomials and "nonsingular" means at these points $\frac{\partial v}{\partial x} \neq 0$ (this is because Hermitian polynomials have simple zeros).

Now because u_λ converges to v in C^1, by the inverse function theorem we know for λ small, near each C_i we have a nonsingular nodal curve of u_λ. In particular, there exist two points $x_{1,\lambda} \neq x_{2,\lambda}$ such that

$$u_\lambda(-1, x_{1,\lambda}) = u_\lambda(-1, x_{2,\lambda}) = 0,$$

if λ small enough. Coming back to u, this means

$$u\left(-\lambda^2, \lambda x_{1,\lambda}\right) = u\left(-\lambda^2, \lambda x_{2,\lambda}\right) = 0.$$

This is a contradiction, because from the previous lemma we know that near $(0,0)$ the nodal curve of u_λ (that is, of u) is a single continuous curve, which implies the existence of a unique $\alpha(-\lambda^2)$ such that

$$u\left(-\lambda^2, \alpha\left(-\lambda^2\right)\right) = 0. \qquad \qquad \square$$

Remark 4.2.9 This result is a generalizations of the Sturm theory for heat equations to our current setting. See also [7] and Sect. 10 in [2].

With this regularity property of the free boundary, we can give an "energy identity". Define the energy

$$E(t) := \sum_i \int_{[0,1]} \frac{1}{2} \left| \frac{\partial u_i}{\partial x} \right|^2 - \frac{1}{2} a_i u_i^2 + \frac{1}{3} u_i^3.$$

We have

Proposition 4.2.10 *For* $t \in (T_{i+1}, T_i)$

$$\frac{d}{dt} E(t) = -\sum_i \int_{[0,1]} \left| \frac{\partial u_i}{\partial t} \right|^2. \qquad (4.11)$$

In particular, $E(t)$ is decreasing in t.

Proof We have

$$\frac{d}{dt} \int_{\alpha_i(t)}^{\beta_i(t)} \frac{1}{2} \left| \frac{\partial u_i}{\partial x} \right|^2 - \frac{1}{2} a_i u_i^2 + \frac{1}{3} u_i^3$$

$$= \int_{\alpha_i(t)}^{\beta_i(t)} \left(\frac{\partial u_i}{\partial x} \frac{\partial^2 u_i}{\partial x \partial t} - a_i u_i \frac{\partial u_i}{\partial t} + u_i^2 \frac{\partial u_i}{\partial t} \right) dt$$

$$- \left[\frac{1}{2} \left(\frac{\partial u_i}{\partial x} \right)^2 - \frac{1}{2} u_i^2 + \frac{1}{3} u_i^3 \right] (\alpha_i(t), t) \frac{d\alpha_i(t)}{dt}$$

$$+ \left[\frac{1}{2} \left(\frac{\partial u_i}{\partial x} \right)^2 - \frac{1}{2} u_i^2 + \frac{1}{3} u_i^3 \right] (\beta_i(t), t) \frac{d\beta_i(t)}{dt}$$

$$= - \int_{\alpha_i(t)}^{\beta_i(t)} \frac{\partial u_i}{\partial t} \left[\frac{\partial^2 u_i}{\partial x^2} + a_i u_i - u_i^2 \right] + \frac{\partial u_i}{\partial x} (\beta_i(t), t) \frac{\partial u_i}{\partial t} (\beta_i(t), t)$$

$$- \frac{\partial u_i}{\partial x} (\alpha_i(t), t) \frac{\partial u_i}{\partial t} (\alpha_i(t), t) - \frac{1}{2} \left| \frac{\partial u_i}{\partial x} \right|^2 (\alpha_i(t), t) \frac{d\alpha_i(t)}{dt}$$

$$+ \frac{1}{2} \left| \frac{\partial u_i}{\partial x} \right|^2 (\beta_i(t), t) \frac{d\beta_i(t)}{dt}. \tag{4.12}$$

By differentiating $u(\alpha_i(t), t) = 0$ and noting the definition of u we have

$$\frac{\partial u_i}{\partial t} (\alpha_i(t), t) + \frac{\partial u_i}{\partial x} (\alpha_i(t), t) \frac{d\alpha_i(t)}{dt} = 0.$$

So we get

$$\frac{\partial u_i}{\partial x} \frac{\partial u_i}{\partial t} (\beta_i(t), t) = - \left| \frac{\partial u_i}{\partial x} \right|^2 (\beta_i(t), t) \frac{d\beta_i(t)}{dt},$$

and

$$\frac{\partial u_i}{\partial x} \frac{\partial u_i}{\partial t} (\alpha_i(t), t) = - \left| \frac{\partial u_i}{\partial x} \right|^2 (\alpha_i(t), t) \frac{d\alpha_i(t)}{dt}.$$

Note that we have $\alpha_i(t) = \beta_{i-1}(t)$, so summing (4.12) in i we get the result.   □

The fourth part of Theorem 4.2.1 can be proved using the same method in [22]. The main idea is that if the solution is very close to a stationary solution for some time, then it either stays near this stationary solution for all the time, or it leaves away and never comes back. Here we will establish a property, which roughly speaking, can be said to be "No extinction at infinity".

From the discussion above, we know for a solution of (4.6), there exists a $T_1 > 0$, such that for all $t > T_1$, no extinction occurs. But at this stage we cannot exclude the possibility that some $u_i \to 0$ as $t \to +\infty$ while remaining nonzero in any finite time. Now let us consider this problem in the following proposition.

Proposition 4.2.11 *For each i, $u_i(x, t)$ converges to $u_i(x)$. If $(u_1(x), u_2(x), \ldots)$ are not identically zero, then each $u_i(x)$ is not identically zero.*

Proof From the energy identity (4.11), we know

$$\lim_{t \to +\infty} \int_t^{+\infty} \int_0^1 \left| \frac{\partial u}{\partial t} \right|^2 = 0.$$

Because

$$\left(\frac{\partial}{\partial t} - \frac{\partial^2}{\partial x^2} \right) \frac{\partial u}{\partial t} = \hat{V} \frac{\partial u}{\partial t},$$

where $\hat{V} := (a_1 - 2u_1) - (a_2 - 2u_2) + \cdots$ is uniformly bounded, by the standard parabolic estimate we get

$$\lim_{t \to +\infty} \sup_x \left| \frac{\partial u}{\partial t} (x, t) \right| = 0.$$

So we have at time t

$$\sup_x \left| \frac{\partial^2 u_i}{\partial x^2} + a_i u_i - u_i^2 \right| \leq \varepsilon(t), \tag{4.13}$$

where $\lim_{t \to +\infty} \varepsilon(t) = 0$. From the statement about the Lipschitz regularity of solutions of (4.6) (for a proof see [2]), $|\frac{\partial u_i}{\partial x}|$ are uniformly bounded. We can multiply (4.13) with it to get

$$\sup_x \left| \frac{\partial^2 u_i}{\partial x^2} \frac{\partial u_i}{\partial x} + a_i u_i \frac{\partial u_i}{\partial x} - u_i^2 \frac{\partial u_i}{\partial x} \right| \leq C\varepsilon(t),$$

for some constant C independent of t. Integrating this in x we get, for $x \in (\alpha_i(t), \beta_i(t))$

$$\left| \left[\frac{1}{2} \left(\frac{\partial u_i}{\partial x} \right)^2 + \frac{1}{2} a_i u_i^2 - \frac{1}{3} u_i^3 \right](x, t) - \left[\frac{1}{2} \left(\frac{\partial u_i}{\partial x} \right)^2 + \frac{1}{2} a_i u_i^2 - \frac{1}{3} u_i^3 \right](\alpha_i(t), t) \right|$$
$$\leq C\varepsilon(t),$$

where $\alpha_i(t)$ can be replaced by $\alpha_{i+1}(t)$. From this, if we define

$$C(t) := \left[\frac{1}{2} \left(\frac{\partial u_i}{\partial x} \right)^2 + \frac{1}{2} a_i u_i^2 - \frac{1}{3} u_i^3 \right](\alpha_i(t), t)$$

for some fixed i, then there exists a constant $C > 0$ such that for all i (similar to the elliptic case considered in Sect. 4.1)

$$\left| \left[\frac{1}{2} \left(\frac{\partial u_i}{\partial x} \right)^2 + \frac{1}{2} a_i u_i^2 - \frac{1}{3} u_i^3 \right](x, t) - C(t) \right| \leq C\varepsilon(t).$$

In particular, if $C(t)$ is uniformly bounded from below by a fixed positive constant, then at time t, if we are at the maximal point of $u_i(x, t)$, because

$$\frac{\partial u_i}{\partial x} = 0,$$

we have

$$\left| \left[\frac{1}{2} a_i u_i^2 - \frac{1}{3} u_i^3 \right](x, t) - C(t) \right| \leq C\varepsilon(t).$$

We can solve this equation to obtain

$$\left| \max_x u_i(x, t) - C_i(t) \right| \leq C\varepsilon(t),$$

where $C_i(t)$ are bounded from below because they only depend on $C(t)$ and a_i. So no $u_i(x, t)$ can converge to 0 for any $t \to +\infty$.

On the other hand, if $\liminf_{t \to +\infty} C(t) = 0$, then we have a sequence $t_k \to +\infty$ such that $\lim_{k \to +\infty} u(x, t_k) = 0$. Moreover, from the standard parabolic estimate, this convergence can be taken in the C^1 topology, and we have for this sequence

$$\lim_{t_k \to +\infty} \sum_i \int_\Omega \frac{1}{2} \left| \frac{\partial u_i}{\partial x} \right|^2 - \frac{1}{2} a_i u_i^2 + \frac{1}{3} u_i^3 = 0.$$

From the energy decreasing property (4.11), we then have

$$\lim_{t \to +\infty} \sum_i \int_\Omega \frac{1}{2} \left| \frac{\partial u_i}{\partial x} \right|^2 - \frac{1}{2} a_i u_i^2 + \frac{1}{3} u_i^3 = 0.$$

In this case, we must have for all i

$$\lim_{t \to +\infty} u_i(x, t) = 0.$$

This is because, any nontrivial solution of the stationary equation (4.1) has energy <0 strictly, which can be seen by an integration by parts. $\square$

The above method can be used to describe connecting orbits, which is defined for all time $t \in (-\infty, +\infty)$. First, from the energy decreasing property we have either

$$\lim_{t \to -\infty} E(t) = +\infty,$$

or

$$\lim_{t \to -\infty} E(t) < +\infty.$$

If the later case happens, the limit as $t \to -\infty$ exists, that is

Proposition 4.2.12 *For a solution u of (4.6) defined in $[0, 1] \times (-\infty, +\infty)$, if*

$$\lim_{t \to -\infty} E(t) < +\infty,$$

then there exists a stationary solution v of (4.6) (that is, a solution of (4.1)), such that

$$\lim_{t \to -\infty} u_i(t) = v_i.$$

Proof From the energy identity, we have

$$\int_{-\infty}^0 \int_{[0,1]} \sum_i \left| \frac{\partial u_i}{dt} \right|^2 < +\infty.$$

In particular,

$$\lim_{t \to -\infty} \int_{-\infty}^{t} \int_{[0,1]} \sum_i \left| \frac{\partial u_i}{dt} \right|^2 = 0.$$

Because $u_i(t)$ are uniformly Lipschitz continuous functions on $[0, 1]$ (note here we have a control of the energy, see [2]), for any sequence $t_k \to -\infty$, there exists a subsequence converging to a Lipschitz continuous function v_i for all i.

We claim that (v_i) is a solution of (4.1). In fact, define

$$u_{i,k}(x, t) = u_i(x, t + t_k).$$

It is a solution of (4.6) defined on $[0, 1] \times (-\infty, +\infty)$ and uniformly Lipschitz continuous, so it converges locally uniformly to a solution v of (4.6). Also we have $u_{i,k}(x, 0) \to v_i(x)$, so

$$v_i(x, 0) = v_i(x).$$

Moreover, we have

$$E(u_k, t) = E(u, t + t_k).$$

So for $t > 0$

$$E(u_k, -t) \le E(u_k, t) + \varepsilon(k, t), \tag{4.14}$$

where $\lim_{k \to +\infty} \varepsilon(k, t) = 0$ for any fixed t. By the uniform Lipschitz continuity, $\forall t$

$$\lim_{k \to +\infty} E(u_k, t) = E(v, t).$$

Passing to the limit in the inequality (4.14) and noting the energy decreasing property (4.11) (this can be applied to the current case), we get

$$E(v, -t) = E(v, t).$$

Using the energy identity (4.11), we have

$$\sum_i \int_{-t}^{t} \int_{[0,1]} \left| \frac{\partial v_i}{\partial t} \right|^2 = 0.$$

This means

$$\frac{\partial v_i}{\partial t} \equiv 0.$$

That is, v_i is the stationary solution of (4.6).

We know that the stationary solutions of (4.6) is a finite set (Theorem 4.1.4). From standard theory, we also know that the ω-limit set of $u(x, t)$ as $t \to -\infty$ is a connected set, so it must be a single point. More precisely, if there exist two limit points, then we can prove the limit set contains a path connecting these two points, contradicting our previous claim. The complete argument is as follows: Take small neighborhoods V_1 and V_2 of the two limiting points v_1, v_2, and neighborhoods V_i of all other stationary solutions v_i, which are disjoint. We know there exist $t_{k,1}$ and $t_{k,2}$ which converge to $-\infty$, such that

$$u(t_{k,1}) \in V_1, \quad \text{and} \quad u(t_{k,2}) \in V_2.$$

From connectedness of the orbit $u(t)$, we know there exists t_k lying between $t_{k,1}$ and $t_{k,2}$, such that

$$u(t_k) \quad \text{is not in} \quad \bigcup_i V_i.$$

But from the uniform Lipschitz continuity, $u(t_k)$ converge to a limit in $C[0, 1]$, which is not a stationary point. This is a contradiction. $\qquad\square$

Remark 4.2.13 For any solution w of (4.6) which exists on $[0, 1] \times (-\infty, +\infty)$, if there exist u and v, which are solutions of (4.1) such that

$$\lim_{t \to -\infty} w_i(x, t) = u_i(x),$$

$$\lim_{t \to +\infty} w_i(x, t) = v_i(x).$$

Then we have two cases:

1. $u \equiv 0$ and v is nontrivial;
2. Both u and v are nontrivial, with the energy of u strictly larger than v's.

Now we consider the uniqueness of the initial value problem (4.6) in a special case.

Proposition 4.2.14 *For the system* (4.6), *if the total number of connected components of the initial value ϕ is finite, then there exists a unique solution.*

Proof The existence is obvious, that is, the solution can be constructed as the limit of the solutions of system (1.1) ($\kappa \to +\infty$). We just need to prove the local (in time) uniqueness.

Assume there exist two local solutions of (4.6), u_i and v_i. By Theorem 4.2.1, we know there exists some $\varepsilon > 0$, such that in $[0, \varepsilon)$, the sum in i of the number of connected components of the support of the component u_i (or v_i) equals that of $\phi = (\phi_1, \phi_2, \ldots, \phi_M)$, and the free boundaries are non-singular curves (see Proposition 4.2.8).

Let

$$d(u, v) = \sum_i |u_i - v_i|.$$

Using the Kato inequality

$$\nabla|h| = sgn(h)\nabla h, \quad \text{a.e.,} \quad \Delta|h| \geq sgn(h)\Delta h, \quad \text{a.e.}$$

we have

$$\left(\frac{\partial}{\partial t} - \frac{\partial^2}{\partial x^2}\right) d(u, v) \leq \sum_i sgn(u_i - v_i)\left(\frac{\partial}{\partial t} - \frac{\partial^2}{\partial^2 x}\right)(u_i - v_i)$$

$$\leq \sum_i sgn(u_i - v_i)\big[a_i - (u_i + v_i)\big](u_i - v_i)$$

$$\leq Cd(u, v).$$

Here $sgn(u_i - v_i)$ is the signature of $u_i - v_i$ and C is a constant depending only on a_i and ϕ. The second inequality is valid by the following argument: by the regularity of the free boundaries, in fact we have

$$\left(\frac{\partial}{\partial t} - \frac{\partial^2}{\partial x^2}\right) u_i = a_i u_i - u_i^2 + \sum_{\alpha_i(t)} \frac{\partial u_i}{\partial v} \delta_{\alpha_i(t)},$$

where $\alpha_i(t)$ is the boundary of $\{u_i > 0\}$ at time t (it consists of finitely many points) and v is the outward unit normal vector to $\partial\{u_i > 0\}$ (although in 1 dimension this is trivial, we keep this notation for clearness), and δ is the Dirac measure supported on these points. Summing these terms in i we find that in the interior, these δ measures cancel each other and we are left with the regular part; while those terms at boundary points 0 (or 1) are of the form

$$sgn(u_i - v_i)\frac{\partial(u_i - v_i)}{\partial v}\delta_0.$$

If near $x = 0$, $u_i - v_i > 0$, then $\frac{\partial(u_i - v_i)}{\partial v}(0) \leq 0$ and vice versa. So this is a non-positive term and in the inequality we can throw it away.

Now $d(u, v) \equiv 0$ at $t = 0$, then the maximum principle gives

$$d(u, v) \equiv 0$$

for all $t \in [0, \varepsilon)$, that is

$$u_i \equiv v_i. \qquad \qquad \square$$

Chapter 5
Approximate Clean Up Lemma

Abstract In this chapter, we establish some technical results, mainly an "Approximate Clean Up lemma" (valid in arbitrary dimension, which allows us to continue to consider the problems in higher dimensions in the future) and some of its consequences. This lemma, in some sense, describes the little invading property of strongly competing system. For the original Clean Up lemma, see Caffarelli et al. (J. Fixed Point Theory Appl. 5(2), 319–351, 2009). Since these results are only intended for the application to the main result in Chap. 6, and the proof is rather technical, at the first reading the readers need only know the conclusions and directly go to Chap. 6, maybe finally come back to read the details of the proof. We mainly consider the following simplified model.

$$\frac{\partial u_i}{\partial t} - \Delta u_i = -\kappa u_i \sum_{j \neq i} u_j. \tag{5.1}$$

The original problem (1.1) can be treated with small changes, which we will indicate in Sect. 5.1. This is because (1.1) can be seen as a perturbation of (5.1). The proof of the Approximate Clean Up Lemma follows the iteration scheme used in Caffarelli et al. (J. Fixed Point Theory Appl. 5(2), 319–351, 2009). After establishing this lemma, we also give a linearization version of this lemma (Corollary 5.3.2 and Proposition 5.3.3), by establishing a lower bound for the sum of the two dominating species near the regular part of the free boundaries (Proposition 5.3.1). Finally, we also include a boundary version of the Approximate Clean Up lemma. In this chapter, by saying that a quantity $\varepsilon(\kappa)$ (depending on κ) converges to 0 rapidly, we mean, for some $\alpha > 0$,

$$\varepsilon(\kappa) \le e^{-\kappa^\alpha}.$$

5.1 Systems with Zeroth Order Terms

In the paper [2], a Clean Up lemma was established (see their Sect. 3, Theorem 11) for the following system on a domain in $\mathbb{R}^n$:

K. Wang, *Free Boundary Problems and Asymptotic Behavior of Singularly Perturbed Partial Differential Equations*, Springer Theses, DOI 10.1007/978-3-642-33696-6_5, © Springer-Verlag Berlin Heidelberg 2013

$$\begin{cases} \Delta u_i \geq 0, \\ \Delta\left(u_i - \sum_{j \neq i} u_j\right) \leq 0, \\ u_i \geq 0, \\ u_i u_j = 0, \quad \text{for } i \neq j. \end{cases} \tag{5.2}$$

We need to consider a system with zeroth order perturbation. Assume we are in the unit ball $B_1(0) \subset \mathbb{R}^n$.

$$\begin{cases} -\Delta u_i \leq f_i(u_i), \\ -\Delta\left(u_i - \sum_{j \neq i} u_j\right) \geq f_i(u_i) - \sum_{j \neq i} f_j(u_j), \\ u_i \geq 0, \\ u_i u_j = 0, \quad \text{for } i \neq j. \end{cases} \tag{5.3}$$

Here f_i, $1 \leq i \leq M$, are given Lipschitz continuous function defined on $\mathbb{R}^+$ with $f_i(u) = a_i u - u^2$ (we can allow more general nonlinearity, which we do not pursue here). Then we have the following "Clean Up Lemma".

Theorem 5.1.1 *Assume at $0 \in B_1(0)$, there exists a sequence $\lambda_k \to 0$ such that the vector function $\frac{1}{\lambda_k} u(\lambda_k x)$ converges to*

$$\hat{u}_1 = \alpha x_1^+, \qquad \hat{u}_2 = \alpha x_1^-, \qquad \hat{u}_j = 0 \quad \text{for } j > 2.$$

Then in a neighborhood of 0

$$\sum_{j > 2} u_j \equiv 0.$$

Remark 5.1.2 It is easy to see that the above blow up limit satisfies the system (5.2) on $\mathbb{R}^n$ (using the fact $f_i(0) = 0$). The proof of this theorem is almost the same to the one in [2], because, after restricting to a small ball and rescaling, (5.3) can be seen as a small perturbation of (5.2).

Now we give the proof of this theorem. We only show the necessary modifications, and for more details please see the proof of the Clean Up lemma in [2].

First by rescaling $\hat{u}(x) = \frac{1}{R} u(Rx)$, we get

$$-\Delta \hat{u}_i(x) = -R \cdot \Delta u_i(Rx) \leq R f_i\big(R\hat{u}_i(x)\big).$$

Thus, by letting R sufficiently small, we can assume u is defined on B_1 with the Lipschitz constant of all f_i smaller than a small $\theta > 0$ (which will be determined later).

By the assumption of Theorem 5.1.1, we can also assume that in $B_1(0)$

$$\left|u_1 - x_1^+\right| \le h_0, \qquad \left|u_2 - x_1^-\right| \le h_0, \qquad \sum_{j>2} u_j \le h_0,$$

with $h_0 > 0$ small. Then the iteration is defined as (with $R_0 = 1$)

$$\begin{cases} h_{l+1} = h_l^2, \\ R_l = R_{l-1} - h_l^{\frac{1}{2}}. \end{cases}$$

In each ball B_{R_l}, we decompose $u_1 - u_2 = v_l + w_l$, such that v_l satisfies

$$\begin{cases} -\Delta v_l = f_1\left(v_l^+\right) - f_2\left(v_l^-\right), & \text{in } B_{R_l} \\ v_l = u_1 - u_2, & \text{on } \partial B_{R_l}. \end{cases} \tag{5.4}$$

This decomposition is possible because the solution of this equation is unique. In fact, if there exist two solutions v and $\hat{v}$, then

$$\begin{cases} -\Delta(v - \hat{v}) = c(x)(v - \hat{v}), & \text{in } B_{R_l} \\ v - \hat{v} = 0, & \text{on } \partial B_{R_l}, \end{cases}$$

where for $f(v) = f_1(v^+) - f_2(v^-)$, if $v(x) \ne \hat{v}(x)$

$$c(x) := \frac{f(v(x)) - f(\hat{v}(x))}{v(x) - \hat{v}(x)};$$

and if $v(x) = \hat{v}(x)$, we simply take $c(x) = 0$.

Note that $\|c\|_{L^\infty} \le \theta$ (the Lipschitz constant of f_i). Then $\exists \lambda > 0$, $\forall \varphi \in H_0^1(B_1)$,

$$\int_{B_1} |\nabla \varphi|^2 - c(x)\varphi^2 \ge \lambda \int_{B_1} \varphi^2. \tag{5.5}$$

From this, we easily see that $v \equiv \hat{v}$, that is, the solution of (5.4) is unique.

This inequality also implies, maybe with a different λ, $\forall \varphi \in H_0^1(B_1)$

$$\int_{B_1} |\nabla \varphi|^2 - c(x)\varphi^2 \ge \lambda \int_{B_1} |\nabla \varphi|^2. \tag{5.6}$$

With these preliminaries, we have the following lemmas. The first one is an easy consequence of the weak maximum principle.

Lemma 5.1.3 *In a ball B, for $\varphi \in H^1(B)$ satisfying*

$$\begin{cases} -\Delta \varphi - c\varphi \ge 0, & \text{in } B \\ \varphi \ge 0, & \text{on } \partial B, \end{cases}$$

weakly, then

$$\varphi \geq 0$$

in B.

Proof Multiply the equation with $\varphi^- := \max\{-\varphi, 0\}$, integrate by parts, and then use the inequality (5.5). $\qquad\square$

Next, we need to construct a comparison function.

Lemma 5.1.4 *In a ball B with radius $R \in [\frac{1}{2}, 1]$, for $\varphi \in H^1(B)$ satisfying*

$$\begin{cases} -\Delta\varphi - c\varphi = h, & in\ B \\ \varphi = h, & on\ \partial B, \end{cases}$$

where h is a positive constant, then in B

$$0 \leq \varphi \leq C(n, \theta)h$$

for a constant $C(n, \theta)$ depending only on n and θ.

Proof That $\varphi \geq 0$ is a consequence of the previous lemma.

By considering $\frac{\varphi}{h}$, we can assume $h = 1$. Then with the help of (5.6), we can use the classical Moser iteration to prove the second claim. $\qquad\square$

Finally, we need a lemma concerning the comparison of the gradients of two solutions (or approximate solutions, which can be treated similarly).

Lemma 5.1.5 *If v_1, v_2 are two solutions of (5.4) in B_1, with*

$$\sup_{B_1} |v_1 - v_2| \leq h,$$

then

$$\sup_{B_{1-h^{\frac{1}{2}}}} |\nabla v_1 - \nabla v_2| \leq Ch^{\frac{1}{2}}.$$

Proof First, differentiating the equation for v_i, $i = 1, 2$, we get

$$-\Delta\nabla v_i = f'(v_i)\nabla v_i.$$

Noting here the special form of $f(v_i)$, we can subtract these two equations to get

$$-\Delta(\nabla v_1 - \nabla v_2) = f'(v_1)(\nabla v_1 - \nabla v_2) + \left[f'(v_1) - f'(v_2)\right]\nabla v_2$$
$$= f'(v_1)(\nabla v_1 - \nabla v_2) + d(x)(v_1 - v_2)\nabla v_2, \qquad (5.7)$$

where

$$d(x) = -2v_1^+ + 2v_1^- + 2v_2^+ - 2v_2^-.$$

Next, take a cut-off function $\eta \equiv 1$ in $B_{1-h^{\frac{1}{2}}}$, $\eta \equiv 0$ outside B_1. We have

$$\int |\nabla v_1 - \nabla v_2|^2 \eta^2 + \Delta(v_1 - v_2)(v_1 - v_2)\eta^2 + 2\eta(v_1 - v_2)(\nabla v_1 - \nabla v_2)\nabla\eta = 0.$$

By our assumptions on $\sup_{B_1} |v_1 - v_2|$, this implies

$$\int |\nabla v_1 - \nabla v_2|^2 \eta^2 \leq \int (C + |\nabla \eta|^2)|v_1 - v_2|^2 \eta^2 \leq Ch.$$

Combining this with (5.7), we can prove our result by standard Moser iteration using (5.6).   □

Noting the fact that if we choose R small enough at the beginning (depending on h_0), then

$$|\Delta x_1 + f_1(x_1^+) - f_2(x_1^-)| \leq h_0, \quad \text{in } B_1.$$

This then implies

$$\sup_{B_1} |v_0 - x_1| \leq Ch_0,$$

$$\sup_{B_{1-h_0^{\frac{1}{2}}}} |\nabla v_0 - \nabla x_1| \leq Ch_0^{\frac{1}{2}},$$

where C depends on n and θ only. The first estimate can be proved by Moser iteration using (5.6) (cf. p. 191 in [25]), and the second estimate can be proved using compactness, analogous to Caffarelli's treatment on perturbation [1], provided at the beginning we have chosen the R small enough. (Assume there exists $h_0 > 0$, such that there exists a sequence of $R_m \to 0$ and a sequence of $v_{m,0}$ constructed as above, but the above estimate is not valid. Then we can pass to the limit to get a contradiction). Now we can give the following decay estimate.

Lemma 5.1.6 *In B_{R_l}, we have*

1. $|v_l - (u_1 - u_2)| \leq h_l$;
2. $|\nabla(v_l - v_{l-1})| \leq h_{l-1}^{\frac{1}{2}}$;
3. $|\nabla v_l - e_1| \leq \sum_{j=1}^{l} h_{j-1}^{\frac{1}{2}} \leq \frac{1}{4}$;
4. *The level surface of v_l is Lipschitz with Lipschitz constant less than* 3.

The proof of the decay of $\sum_{j>2} u_j$ from B_{R_l} to $B_{R_{l+1}}$ is almost the same as in [2], with the only difference, that we now only have

$$\Delta u_i \geq -\theta u_i.$$

But from this we can still get a weaker mean value inequality similar to the one for subharmonic functions.

We also need to show that in $B_{R_{l+1}}$, w_{l+1} can be controlled by $\sum_{j>2} u_j$. This can be seen by the equations they satisfy

$$-\Delta w_{l+1} = \Delta v_{l+1} - \Delta(u_1 - u_2)$$

$$= -f(v_{l+1}) + f(u_1 - u_2) - \sum_{i>2} \mu_{1i} + \sum_{i>2} \mu_{2i}$$

$$= -c(x)w_{l+1} - \sum_{i>2} \mu_{1i} + \sum_{i>2} \mu_{2i},$$

where f and c are defined as before, and for $i \neq j$, μ_{ij} are Radon measure parts of Δu_i supported on $\partial\{u_i > 0\} \cap \partial\{u_j > 0\}$. While for $\sum_{j>2} u_j$ we have

$$-\Delta \sum_{j>2} u_j = \sum_{j>2} f_j(u_j) - \sum_{j>2} (\mu_{1j} + \mu_{2j}).$$

Here the first term of the right-hand side has the order θh_{l+1} on $B_{R_{l+1}}$. So we have

$$-\Delta w_{l+1} - c(x)w_{l+1} \geq -\Delta \sum_{j>2} u_j - \theta h_{l+1},$$

$$-\Delta w_{l+1} - c(x)w_{l+1} \leq \Delta \sum_{j>2} u_j + \theta h_{l+1}.$$

Or in another form

$$(-\Delta - c(x))\left(w_{l+1} - \sum_{j>2} u_j\right) \geq -\theta h_{l+1},$$

$$(-\Delta - c(x))\left(w_{l+1} - \sum_{j>2} u_j\right) \leq \theta h_{l+1}.$$

From the local interior estimate for the elliptic operator $-\Delta - c(x)$ we get a control

$$\sup_{B_{R_{l+1}}} |w_{l+1}| \leq Ch_{l+1},$$

with constant C depending only on the initial data.

5.2 Proof of the Approximate Clean Up Lemma

First, we need a lemma which is the parabolic analogue of Lemma 4.4 in [11]. In the following, we may not distinguish the various constants in the estimates, which are simply denoted by C, or even omitted if no confusion occurs.

Lemma 5.2.1 *Let* $Q_{2R} := B_{2R}(x) \times (t - 4R^2, t + 4R^2)$, *and* $Q_R := B_R(x) \times (t - R^2, t + R^2)$. *Let* A, M *be two positive constants and* u *be a smooth function in* Q_{2R} *such that*

$$
\begin{cases}
\dfrac{\partial u}{\partial t} - \Delta u \le -Mu, & \text{in } Q_{2R} \\[2mm]
u \ge 0, & \text{in } Q_{2R} \\[2mm]
u \le A, & \text{in } Q_{2R}.
\end{cases}
$$

If M *is sufficiently large compared to* R, *then we have*

$$
\sup_{Q_R} u \le C_1 A e^{-C_2 R M^{\frac{1}{2}}},
$$

where C_1, C_2 *are positive constants depending only on the dimension.*

Proof Given two r_1, r_2, take a cut-off function η, such that

$$
\begin{cases}
\eta \equiv 1, & \text{in } Q_{r_1} \\[2mm]
\eta \equiv 0, & \text{outside } Q_{r_2} \\[2mm]
\left| \dfrac{\partial \eta}{\partial t} \right| \le 10(r_2 - r_1)^{-2} \\[2mm]
|\Delta \eta| \le 10(r_2 - r_1)^{-2}.
\end{cases}
$$

First by standard calculation, we have

$$
\frac{\partial u^2}{\partial t} - \Delta u^2 \le -Mu^2.
$$

Multiplying the equation with η and integrating by parts, we get

$$
\iint_{Q_{r_1}} u^2 \le M^{-1} \iint \left(\Delta u^2 - \frac{\partial u^2}{\partial t} \right) \eta \le M^{-1} \iint u^2 \left(\Delta \eta + \frac{\partial \eta}{\partial t} \right)
$$

$$
\le 20M^{-1}(r_2 - r_1)^{-2} \iint_{Q_{r_2}} u^2.
$$

Now divide the interval $[R, 2R]$ into $N + 1$ equal parts with

$$
r_k = \left(1 + \frac{k}{N} \right) R, \quad \text{for } k = 0, 1, \dots, N.
$$

An iteration in k using the above inequality gives

$$
\iint_{Q_R} u^2 \le C \left(\frac{N^2}{R^2 M} \right)^N \iint_{Q_{2R}} u^2.
$$

If $R\sqrt{M} > 100$ (sufficiently large is enough), we can take $N = [\frac{1}{2}R\sqrt{M}]$ (the integer part) to get, for two positive constants C_1, C_2 depending only on the dimension,

$$\iint_{Q_R} u^2 \leq C_1 e^{-C_2 RM^{\frac{1}{2}}} \iint_{Q_{2R}} u^2.$$

Then by the sup bound estimate of the sub-caloric function (Theorem 6.17, p. 121 in [30]), we can finish the proof of this lemma.   □

Theorem 5.2.2 *Assume a sequence of the solution of* (5.1), u_{i,κ_n}, *converge to* u_i. *If we have a point of the free boundaries* (x, t), *where in a neighborhood there are exactly two components* u_1, u_2 *which do not vanish, and* $|\nabla u_1(x, t)| \neq 0$, *then in a small parabolic cylinder* $Q_R(x, t) := B_R(x) \times (t - R^2, t + R^2)$, *we have*

$$\sup_{Q_R} \sum_{i \neq 1,2} u_{i,\kappa_n} \leq C(R) e^{-\kappa^{1/12}},$$

for κ *large, where* $C(R)$ *is independent of* κ.

Proof Assume the free boundary point is $(0, 0)$. After a rescaling (this changes κ by a fixed scale), we may assume we are in Q_1, and here we have

$$\begin{cases} \left(\dfrac{\partial}{\partial t} - \Delta\right)(u_1 - u_2) = 0, \\ \left|\nabla\big(u_1(x, t) - u_2(x, t)\big) - e_1\right| \leq \varepsilon, \\ \nabla\big(u_1(0, 0) - u_2(0, 0)\big) = e_1. \end{cases}$$

Here the unit vector $e_1 = (1, 0, 0, \ldots, 0) \in \mathbb{R}^n$ and ε is a small positive constant. Given a $h_0 > 0$ small enough, which will be chosen later, we know for κ large enough

$$\sup_{Q_1} \left|(u_{1,\kappa} - u_{2,\kappa}) - (u_1 - u_2)\right| \leq h_0,$$

$$\sup_{Q_1} \left|\sum_{i>2} u_{i,\kappa}\right| \leq h_0.$$

Define the iteration (starting with $R_0 = 1$)

$$\begin{cases} h_{l+1} = h_l^2, \\ R_l = R_{l-1} - h_{l-1}^{\frac{1}{2}}. \end{cases}$$

We know if h_0 is small enough, h_l will converge to 0 rapidly, and $\lim_{l \to +\infty} R_l \geq \frac{1}{2}$. We need to estimate the decay of $\sup_{Q_{R_l}} \sum_{i \neq 1,2} u_{i,\kappa}$.

In the cylinder Q_{R_l}, we decompose $u_{1,\kappa} - u_{2,\kappa} := v_l + w_l$, where

$$\begin{cases} \left(\dfrac{\partial}{\partial t} - \Delta\right) v_l = 0, & \text{in } Q_{R_l} \\ v_l = u_{1,\kappa} - u_{2,\kappa}, & \text{on } \partial_p Q_{R_l}. \end{cases}$$

Note here for $l = 0$, because $|v_0 - (u_1 - u_2)| \le h_0$ on $\partial_p Q_1$ and both satisfy the heat equation, we then have

$$\sup_{Q_1} \left| v_0 - (u_{1,\kappa} - u_{2,\kappa}) \right| \le 2h_0.$$

In order to complete the proof, we need the following lemma.

Lemma 5.2.3 *If $h_l \ge \kappa^{-\frac{1}{6}}$, then in Q_{R_l}*

1. $|v_l - (u_{1,\kappa} - u_{2,\kappa})| \le e^{-h_{l-1}^{-\frac{1}{2}}} + h_{l-1} e^{-h_{l-1}^{\frac{3}{2}} \kappa^{\frac{1}{2}}} \le h_{l-1}^2$;
2. $|\nabla(v_l - v_{l-1})| \le h_{l-1}^{\frac{1}{2}}$;
3. $|\nabla v_l - e_1| \le \sum_{j=1}^{l} h_{j-1}^{\frac{1}{2}} \le \frac{1}{4}$;
4. *The level surface of v_l is Lipschitz with Lipschitz constant less than 3.*

We divide the cylinder $Q_{R_{l+1}}$ into two parts: the good part where $|v_l| \ge 2h_l$ and the bad part where $|v_l| \le 2h_l$.

In the first part,

$$u_{1,\kappa} + u_{2,\kappa} \ge |u_{1,\kappa} - u_{2,\kappa}| \ge v_l - h_l \ge h_l.$$

By the third result in Lemma 5.2.3 (combining with a standard a priori estimate of the heat equation, see p. 17 in [30]), v_l is Lipschitz with Lipschitz constant less than 3, so if $v_l(x, t) \ge 2h_l$, there is a cylinder of size $\sim h_l$, such that on this cylinder, $v_l(x, t) \ge \frac{1}{2}h_l$. Then we can use Lemma 5.2.1, to get that

$$\sum_{i \ne 1,2} u_{i,\kappa}(x, t) \le h_l e^{-h_l^{\frac{3}{2}} \sqrt{\kappa}}. \tag{5.8}$$

In the remaining part where $|v_l| \le 2h_l$, note that it is a narrow domain in the sense of [2], Lemma 24. We have the following estimate.

Lemma 5.2.4 *For $(x, t) \in \{|v_l| \le 2h_l\} \cap Q_{R_l}$, we have*

$$\sum_{i \ne 1,2} u_{i,\kappa}(x, t) \le e^{-\frac{R_l - \sqrt{|t| + |x|^2}}{h_l}} h_l + h_l e^{-h_l^{\frac{3}{2}} \sqrt{\kappa}}. \tag{5.9}$$

Proof Take an integer N such that $\lambda N h_l \sim R_l - \sqrt{|t| + |x|^2}$ and define $Q_j :=$ $Q_{R_l - j\lambda h_l}$, $j = 0, 1, \ldots, N$ where λ will be determined later. From $\hat{Q}_j := Q_j \cap$ $\{|v_l| \leq h_l\}$ to $\hat{Q}_{j+1}$ we have the following decay estimate, using the sup norm estimate of sub-caloric function:

$$
\sup_{\hat{Q}_{j+1}} u_{i,\kappa}^2 \leq C(n) \frac{1}{|Q_{\lambda h_l}|} \iint_{Q_{\lambda h_l}} u_{i,\kappa}^2
$$

$$
\leq C(n) \frac{1}{|Q_{\lambda h_l}|} \left[\iint_{Q_{\lambda h_l} \cap \{|v_l| \geq 2h_l\}} u_{i,\kappa}^2 + \iint_{Q_{\lambda h_l} \cap \{|v_l| \leq 2h_l\}} u_{i,\kappa}^2 \right]
$$

$$
\leq C(n) h_l e^{-h_l^{\frac{3}{2}} \sqrt{\kappa}} + \frac{1}{2} \sup_{\hat{Q}_j} u_{i,\kappa}^2,
$$

if $\{|v_l| \leq 2h_l\} \cap Q_{R_l}$ is narrow in the sense that, for any $(x, t) \in \{|v_l| \leq 2h_l\} \cap Q_{R_l}$,

$$
\left| Q_{\lambda h_l}(x, t) \cap \{|v_l| \leq 2h_l\} \cap Q_{R_l} \right| \leq \frac{1}{2C(n)} \left| Q_{\lambda h_l}(x, t) \right|.
$$

We can choose λ, which depends only on the Lipschitz constant of v_l, to satisfy this condition.

Finally, an iteration of the above decay estimate in j gives our result. In fact,

$$
\sup_{\hat{Q}_N} u_{i,\kappa}^2 \leq \left(1 + \frac{1}{2} + \cdots + \frac{1}{2^N} \right) C(n) h_l e^{-h_l^{\frac{3}{2}} \sqrt{\kappa}} + \frac{1}{2^{N+1}} \sup_{\hat{Q}_0} u_{i,\kappa}^2. \qquad \square
$$

Now in $Q_{R_{l+1}}$,

$$
\left(\frac{\partial}{\partial t} - \Delta \right) w_{l+1} = \left(\frac{\partial}{\partial t} - \Delta \right) (u_{1,\kappa} - u_{2,\kappa}) \leq \left(\frac{\partial}{\partial t} - \Delta \right) \sum_{i \neq 1,2} u_{i,\kappa},
$$

and

$$
\left(\frac{\partial}{\partial t} - \Delta \right) w_{l+1} \geq -\left(\frac{\partial}{\partial t} - \Delta \right) \sum_{i \neq 1,2} u_{i,\kappa}. \tag{5.10}
$$

We also have $w_{l+1} = 0$ on $\partial_p Q_{R_{l+1}}$, so in $Q_{R_{l+1}}$

$$
|w_{l+1}| \leq \sum_{i \neq 1,2} u_{i,\kappa} \leq h_l e^{-h_l^{-\frac{1}{2}}} + h_l e^{-h_l^{\frac{3}{2}} \sqrt{\kappa}}. \tag{5.11}
$$

Here the second inequality follows from (5.8) and (5.9). This gives the first part of Lemma 5.2.3.

Now we also have in $Q_{R_{l+1}}$

$$|v_l - v_{l+1}| \leq |v_l - (u_{1,\kappa} - u_{2,\kappa})| + |(u_{1,\kappa} - u_{2,\kappa}) - v_{l+1}| \leq h_l,$$

and an interior gradient estimate gives the second part of Lemma 5.2.3 (after shrinking $Q_{R_{l+1}}$ by a factor $h_l^{\frac{1}{2}}$). The remaining parts are easy consequences of the second part. This finishes the proof of Lemma 5.2.3.

From (5.11), we also get, at the end point of our iteration, where $h_{l_0} = \kappa^{-\frac{1}{6}}$, in $Q_{R_{l_0}}$ with $R_{l_0} \geq \frac{1}{2}$

$$\sum_{i \neq 1,2} u_{i,\kappa} \leq e^{-h_{l_0}^{-\frac{1}{2}}} + h_{l_0} e^{-h_{l_0}^{\frac{3}{2}} \sqrt{\kappa}} \leq e^{-\kappa^{\frac{1}{12}}}.$$

This finishes the proof of Theorem 5.2.2.   □

Remark 5.2.5 From Proposition 4.2.7, for those solutions in Theorem 4.1, we know the condition in Theorem 5.2.2 is satisfied except for finitely many times $T_1, T_2, \ldots$, in particular, in $[0, 1] \times (T_1, +\infty)$.

In 1 dimension, in the case of elliptic equation (a system of ODEs), the iteration scheme in the proof is not needed (in fact, can not be applied). However in this case we can prove the conclusion directly. That is, if we are in the position of the starting point of the iteration scheme, instead of defining an iteration, we estimate $\sum_{i>2} u_{i,\kappa}$ in the two domains where $|u_{1,\kappa} - u_{2,\kappa}| \geq h$ or $|u_{1,\kappa} - u_{2,\kappa}| \leq h$, separately.

In the first domain, we have $\kappa(u_{1,\kappa} + u_{2,\kappa}) \geq \kappa h$, so

$$\sum_{i>2} u_{i,\kappa} \leq e^{-\kappa h^{\frac{3}{2}}}.$$

In the second domain, which we can assume to be $[-h, h]$, u_κ satisfies (here θ is the Lipschitz constant of f_i)

$$-u_{i,\kappa}'' \leq \theta u_{i,\kappa}.$$

We claim that, for some constant $C > 0$

$$\sup_{[-h,h]} u_{i,\kappa} \leq \max\{u_{i,\kappa}(-h), u_{i,\kappa}(h)\}(1 + Ch^2).$$

This can be seen by rescaling $u_{i,\kappa}$ to $\hat{u}_{i,\kappa}$, which is defined on $[-1, 1]$:

$$\hat{u}_{i,\kappa}(x) := \frac{1}{\max\{u_{i,\kappa}(-h), u_{i,\kappa}(h)\}} u_{i,\kappa}(hx)$$

satisfying

$$-\hat{u}_{i,\kappa}'' \leq \theta h^2 \hat{u}_{i,\kappa}.$$

Corollary 5.2.6 *Under the assumption of Theorem 5.2.2, in $Q_{\frac{R}{2}}(x, t)$, we have for* $u_\kappa = u_{1,\kappa} - u_{2,\kappa}$:

$$\left(\frac{\partial}{\partial t} - \Delta\right)u_\kappa = \varepsilon(\kappa),$$

where $\lim_{\kappa \to +\infty} \varepsilon(\kappa) = 0$.

5.3 A Linearized Version

In this and the following section, we study the convergence of the solutions to the linearized equations of (5.1), under the same assumption of the Approximate Clean Up lemma. We need these results in Chap. 6, in order to get a nontrivial solution of the linearized system of the limit system as $\kappa \to +\infty$. (This then contradicts our nondegeneracy assumption, see Chap. 6.)

We have shown that in a neighborhood of a regular point of free boundaries, where in the limit case only u_1 and u_2 are not vanishing, then for $j \neq 1, 2$, $u_{j,\kappa} \leq e^{-\kappa^\alpha}$ for some $\alpha \in (0, 1)$. It is easily seen that here we must have

$$\kappa(u_{1,\kappa} + u_{2,\kappa}) \to +\infty.$$

In the following proposition, we will give an explicit lower bound, which also improve the decay rate in Theorem 5.2.2 a postiori.

Proposition 5.3.1 *Under the assumption of Theorem 5.2.2, there exists a constant C, such that in a small cylinder*

$$\kappa^{\frac{1}{3}}(u_{1,\kappa} + u_{2,\kappa}) \geq C.$$

Proof Note that the elliptic case can be proved similarly. Assume in a cylinder Q_R, $u_{i,\kappa}$ satisfy the equation

$$\frac{\partial u_{i,\kappa}}{\partial t} - \Delta u_{i,\kappa} = -\kappa u_{i,\kappa} \sum_{j \neq i} u_{j,\kappa}, \tag{5.12}$$

and $u_{j,\kappa} \to 0$ rapidly for $j \neq 1, 2$; $u_\kappa = u_{1,\kappa} - u_{2,\kappa}$ satisfies

$$\frac{\partial u_\kappa}{\partial t} - \Delta u_\kappa = \varepsilon(\kappa), \tag{5.13}$$

where $\varepsilon(\kappa) \to 0$ rapidly; $u_{1,\kappa} \to u_1$, and $u_{2,\kappa} \to u_2$ uniformly in Q_R, with $u = u_1 - u_2$ caloric in Q_R, and by the interior gradient estimate of caloric functions,

$$|\nabla u - e| \leq \varepsilon \tag{5.14}$$

for a fixed unit vector e and a small $\varepsilon > 0$ (this is derived from the assumption of Theorem 5.2.2, see the third result in Lemma 5.2.3).

We want to prove that in the cylinder $Q_{\frac{R}{2}}$

$$\kappa^{\frac{1}{3}}(u_{1,\kappa} + u_{2,\kappa}) \geq C.$$

Assume by the contrary, there exist $X_\kappa \in Q_{\frac{R}{2}}$, such that

$$\kappa^{\frac{1}{3}}(u_{1,\kappa} + u_{2,\kappa})(X_\kappa) = \varepsilon(\kappa),$$

where $\lim_{\kappa \to +\infty} \varepsilon(\kappa) = 0$.

Without loss of generality, we can assume $X_\kappa \to X_0$ for some $X_0 \in Q_{\frac{R}{2}}$.

From (5.13), we know u_κ are uniformly bounded, and close to u if κ is large, in $C^{1+\alpha, \frac{\alpha}{2}}(Q_R)$ for any $\alpha \in (0, 1)$. So without loss of generality, we can assume

1. $|\nabla u_\kappa - e| \leq 2\varepsilon$, and $|\nabla u_\kappa| \geq 1 - 2\varepsilon$;
2. by the implicit function theorem (applied to each time sheet $Q_r \cap \{t\}$), $\Sigma_\kappa := \{u_\kappa = 0\}$ can be represented by the graph of a function φ_κ in the direction e, with φ_κ uniformly bounded in $C^{1+\alpha, \frac{\alpha}{2}}(Q_R)$ for some $\alpha \in (0, 1)$;
3. by the lower bound on $|\nabla u_\kappa|$, $|u_\kappa(X)| \geq \frac{1}{2} dist(X, \Sigma_\kappa)$.

Now we have

$$\varepsilon(\kappa)\kappa^{-\frac{1}{3}} \geq (u_{1,\kappa} + u_{2,\kappa})(X_\kappa) \geq |u_\kappa(X_\kappa)|$$

$$\geq \frac{1}{2} dist(X_\kappa, \Sigma_\kappa)$$

$$\geq \frac{1}{2} dist(X_\kappa, Y_\kappa),$$

where $Y_k \in \Sigma_\kappa$ (at the same time slice) satisfies $dist(X_\kappa, Y_\kappa) = dist(X_\kappa, \Sigma_\kappa)$. From the above inequality we know $dist(X_\kappa, Y_\kappa) \to 0$. In particular, $Y_\kappa \to X_0$ too.

Now define the rescaling (here we define the rescaling $\lambda X = \lambda(x, t) := (\lambda x, \lambda^2 t)$)

$$\hat{u}_{i,\kappa}(X) = \kappa^{\frac{1}{3}} u_{i,\kappa}\left(Y_\kappa + \kappa^{-\frac{1}{3}} X\right),$$

which satisfy the equations

$$\frac{\partial \hat{u}_\kappa}{\partial t} - \Delta \hat{u}_{i,\kappa} = \hat{u}_{i,\kappa} \sum_{j \neq i} \hat{u}_{j,\kappa}$$

on $Q_{\kappa^{\frac{1}{3}} R}$.

From the rapid decay of $u_{j,\kappa}$ for $j \neq 1, 2$, we have

$$\sup_{Q_{\kappa^{\frac{1}{3}} R}} \hat{u}_{j,\kappa} \to 0.$$

We also have

$$\hat{X}_\kappa = \kappa^{\frac{1}{3}}(X_\kappa - Y_\kappa) \to 0,$$

and for $i = 1, 2$

$$\hat{u}_{i,\kappa}(\hat{X}_\kappa) \to 0; \tag{5.15}$$

and for $\hat{u}_\kappa = \hat{u}_{1,\kappa} - \hat{u}_{2,\kappa}$

$$|\nabla \hat{u}_\kappa| = |\nabla u_\kappa| \geq \frac{1}{2}. \tag{5.16}$$

From the uniform Lipschitz continuity of $u_{i,\kappa}$ in Chap. 2, we have for some constant C independent of κ

$$|\nabla \hat{u}_{i,\kappa}| = |\nabla u_{i,\kappa}| \leq C.$$

Here we need to note that, the uniform Lipschitz continuity of $u_{i,\kappa}$ was proven in Chap. 2 only in the symmetric case, that is, $\forall i, j, b_{ij} = b_{ji}$. If b_{ij} is not symmetric, we can use Theorem 5.2 in [11]. That theorem is proven for the case of two equations. Here by noting (5.12) and (5.13), locally our system can be reduced to two equations about $u_{1,\kappa}$ and $u_{2,\kappa}$, with error terms exponentially small, so that theorem can still be applied.

Now it is easily seen that

$$\frac{\partial \hat{u}_{1,\kappa}}{\partial t} - \Delta \hat{u}_{1,\kappa} = -\hat{u}_{1,\kappa} \hat{u}_{2,\kappa} + \varepsilon(\kappa);$$

$$\frac{\partial \hat{u}_{2,\kappa}}{\partial t} - \Delta \hat{u}_{2,\kappa} = -\hat{u}_{1,\kappa} \hat{u}_{2,\kappa} + \varepsilon(\kappa).$$

This means

$$\frac{\partial \hat{u}_\kappa}{\partial t} - \Delta \hat{u}_\kappa = \varepsilon(\kappa).$$

Combing these three equations with (5.15), and noting the uniform Lipschitz continuity, we get

1. $\hat{u}_{i,\kappa} \to \hat{u}_i$ and $\nabla \hat{u}_{i,\kappa} \to \nabla \hat{u}_i$ uniformly on any compact set of $\mathbb{R}^n \times (-\infty, +\infty)$ (for κ large enough), $i = 1, 2$, with

$$\frac{\partial \hat{u}_1}{\partial t} - \Delta \hat{u}_1 = \frac{\partial \hat{u}_2}{\partial t} - \Delta \hat{u}_2 = -\hat{u}_1 \hat{u}_2;$$

2. $\hat{u}_\kappa \to \hat{u}$ locally uniformly with

$$\frac{\partial \hat{u}}{\partial t} - \Delta \hat{u} = 0.$$

From (5.15), we get

$$\hat{u}_1(0) + \hat{u}_2(0) = \lim_{\kappa \to +\infty} \hat{u}_{1,\kappa}(\hat{x}_\kappa) + \hat{u}_{2,\kappa}(\hat{x}_\kappa) = 0.$$

We also have $\hat{u}_1, \hat{u}_2 \geq 0$, so by the strong maximum principle we get

$$\hat{u}_1 \equiv \hat{u}_2 \equiv 0.$$

In particular

$$\hat{u} \equiv 0.$$

But by taking limit in (5.16), we also have

$$|\nabla \hat{u}| \geq \frac{1}{2},$$

which is a contradiction, and our claim is proven. $\square$

With the help of this proposition, we have the following result.

Corollary 5.3.2 *Under the assumption of Theorem 5.2.2, and if in $Q_R(x, t)$, we have for $f_{i,\kappa}$, which is uniformly bounded independent of κ, satisfy:*

$$\left(\frac{\partial}{\partial t} - \Delta\right) f_{i,\kappa} = -\kappa f_{i,\kappa} \sum_{j \neq i} u_{j,\kappa} - \kappa u_{i,\kappa} \sum_{j \neq i} f_{j,\kappa}.$$

Then in $Q_{\frac{R}{2}}(x, t)$

$$f_{j,\kappa} \to 0 \text{ rapidly}, \quad \text{for } j \neq 1, 2.$$

Moreover,

$$\left(\frac{\partial}{\partial t} - \Delta\right)(f_{1,\kappa} - f_{2,\kappa}) = \varepsilon(\kappa),$$

where $\lim_{\kappa \to +\infty} \varepsilon(\kappa) = 0$ rapidly.

Proof For $j \neq 1, 2$, we have

$$\left(\frac{\partial}{\partial t} - \Delta\right) f_{j,\kappa} = -\kappa f_{j,\kappa}(u_{1,\kappa} + u_{2,\kappa}) + \varepsilon(\kappa).$$

Then because there exists C such that $\sup |f_{j,\kappa}| \leq C$, we have

$$\left(\frac{\partial}{\partial t} - \Delta\right)(f_{j,\kappa})^2 \leq -\kappa (f_{j,\kappa})^2 (u_{1,\kappa} + u_{2,\kappa}) + \varepsilon(\kappa). \tag{5.17}$$

Because $\kappa(u_{1,\kappa} + u_{2,\kappa}) \geq \kappa^{\frac{2}{3}}$ in a fixed small cylinder, we can proceed as in the proof of Lemma 5.2.1, to prove that $(f_{j,\kappa})^2$ converge to 0 rapidly in a small cylinder. This is a local result, and a procedure by covering gives the global result. With this estimate in hand, we see

$$\left(\frac{\partial}{\partial t} - \Delta\right)(f_{1,\kappa} - f_{2,\kappa}) = -\kappa(f_{1,\kappa} - f_{2,\kappa}) \sum_{j \neq 1,2} u_{j,\kappa} - \kappa(u_{1,\kappa} - u_{2,\kappa}) \sum_{j \neq 1,2} f_{j,\kappa},$$

and we can conclude the claim. $\square$

The assumption in this corollary on the sup norm is not so satisfactory. Instead we give the following proposition.

Proposition 5.3.3 *Under the same assumptions as in the previous corollary, except that we only assume $\int_{Q_R} \sum_i |f_i|^q$, for some $q > 2(n+3)$, are uniformly bounded, we still have the same conclusion.*

Proof First, for $f_{i,\kappa}^2$, $i > 2$, it satisfies

$$\left(\frac{\partial}{\partial t} - \Delta\right) f_{i,\kappa}^2 \leq 2 f_{i,\kappa} \left(\frac{\partial}{\partial t} - \Delta\right) f_{i,\kappa} \leq -\kappa f_{i,\kappa}^2 \sum_{j\neq i} u_{j,\kappa} - \kappa u_{i,\kappa} \sum_{j\neq i} f_{j,\kappa} f_{i,\kappa},$$

where, by Proposition 5.3.1

$$\kappa \sum_{j\neq i} u_{j,\kappa} \geq \kappa(u_{1,\kappa} + u_{2,\kappa}) \geq \kappa^{\frac{2}{3}}.$$

We can proceed as in the proof of Lemma 5.2.1, multiplying the equation with η and integrating by parts to get the following. (Simply denote $f_{i,\kappa}$ by f, $M := \kappa^{\frac{2}{3}}$ and $g := \kappa u_{i,\kappa} \sum_{j\neq i} f_{j,\kappa} f_{i,\kappa}$.)

$$\iint_{Q_{\frac{r_1+r_2}{2}}} f^2 \leq M^{-1} \iint \left(\Delta f^2 - \frac{\partial f^2}{\partial t}\right) \eta + g\eta$$

$$\leq M^{-1} \iint f^2 \left(\Delta \eta + \frac{\partial \eta}{\partial t}\right) + g\eta$$

$$\leq M^{-1} \left\{ 20(r_2 - r_1)^{-2} \iint_{Q_{r_2}} f^2 + \iint |g| \right\}.$$

Define $R_k = \frac{R}{2}(2 - \frac{k}{N})$ for $k = 0, 1, \ldots, N$. We use the above estimate to iterate. Then if M is large enough comparing to R (exactly as in Lemma 5.2.1) we get, for $i > 2$

$$\iint_{Q_{\frac{R}{2}}} f_{i,\kappa}^2 \leq e^{-c(n)\kappa^{\frac{1}{3}} R^2} \iint_{Q_R} f_{i,\kappa}^2 + C(n) \iint_{Q_R} \kappa u_{i,\kappa} \sum_{j\neq i} |f_{j,\kappa}| |f_{i,\kappa}|$$

$$\leq e^{-\kappa^{\alpha}},$$

for a small $\alpha > 0$, because in Q_R, $u_{i,\kappa} \leq e^{-\kappa^{\beta}}$ for some $\beta > 0$.

By the uniform boundedness of $\iint_{Q_R} |f_{i,\kappa}|^q$, and with the help of Hölder inequality, this implies, $\forall 2 < p < q$, $\exists \alpha(p)$ such that

$$\iint_{Q_{\frac{R}{2}}} |f_{i,\kappa}|^p \leq e^{-\kappa^{\alpha(p)}}.$$

Now in $Q_{\frac{R}{2}}$

$$\left(\frac{\partial}{\partial t} - \Delta\right) f_{i,\kappa}^2 \le -\kappa u_{i,\kappa} \sum_{j\ne i} f_{j,\kappa} f_{i,\kappa}.$$

Here we can use the standard De Giorgi estimate (see [29]) to get

$$\sup_{Q_{\frac{R}{4}}} f_{i,\kappa}^2 \le C(n)\left[\left(R^{-n-2}\int_{Q_{\frac{R}{2}}}|f_{i,\kappa}|^4\right)^{\frac{1}{n+3}}\right.$$

$$\left.+ \left(R^{-n-2}\int_{Q_{\frac{R}{2}}}\left|\kappa u_{i,\kappa}\sum_{j\ne i}f_{j,\kappa}f_{i,\kappa}\right|^{n+3}\right)^{\frac{1}{2}}\right]$$

$$\le e^{-\kappa^\alpha},$$

for another $\alpha > 0$.

Next, let us consider $f_{i,\kappa}$ for $i = 1, 2$, in $Q_{\frac{R}{4}}$. By the Kato inequality

$$\begin{cases}\left(\frac{\partial}{\partial t} - \Delta\right)|f_{1,\kappa}| \le -\kappa|f_{1,\kappa}|u_{2,\kappa} + \kappa u_{1,\kappa}|f_{2,\kappa}| + \varepsilon(\kappa),\\[2mm]\left(\frac{\partial}{\partial t} - \Delta\right)|f_{2,\kappa}| \le -\kappa|f_{2,\kappa}|u_{1,\kappa} + \kappa u_{2,\kappa}|f_{1,\kappa}| + \varepsilon(\kappa).\end{cases}$$

Summing these two equations we get

$$\left(\frac{\partial}{\partial t} - \Delta\right)\left(|f_{1,\kappa}| + |f_{2,\kappa}|\right) \le \varepsilon(\kappa).$$

By standard parabolic estimate, $f_{1,\kappa}$ and $f_{2,\kappa}$ are uniformly bounded in $Q_{\frac{R}{8}}$. The rest of the proof is the same as the previous corollary. $\qquad\square$

Remark 5.3.4 If the equation has the form

$$\left(\frac{\partial}{\partial t} - \Delta\right)f_{i,\kappa} = -\kappa f_{i,\kappa}\sum_{j\ne i}b_{ij}u_{j,\kappa} - \kappa u_{i,\kappa}\sum_{j\ne i}b_{ij}f_{j,\kappa}.$$

The proof is also valid. We only need to make some changes in the place where we used the Kato inequality. We modify $f_{1,\kappa}$ and $f_{2,\kappa}$ into

$$\begin{cases}\hat{f}_{1,\kappa} := b_{21}f_{1,\kappa},\\[2mm]\hat{f}_{2,\kappa} := b_{12}f_{2,\kappa},\end{cases}$$

which satisfy

$$\begin{cases} \left(\dfrac{\partial}{\partial t} - \Delta\right) \hat{f}_{1,\kappa} = -\kappa b_{12} \hat{f}_{1,\kappa} u_{2,\kappa} - \kappa b_{21} u_{1,\kappa} \hat{f}_{2,\kappa} + \varepsilon(\kappa), \\[2mm] \left(\dfrac{\partial}{\partial t} - \Delta\right) \hat{f}_{2,\kappa} = -\kappa b_{21} \hat{f_2}, \kappa u_{1,\kappa} - \kappa b_{12} u_{2,\kappa} \hat{f}_{1,\kappa} + \varepsilon(\kappa). \end{cases}$$

Then we can use the Kato inequality again.

5.4 A Boundary Version

In this section, we first consider

$$\left(\frac{\partial}{\partial t} - \Delta\right) u_i = f_i(u_i) - \kappa u_i \sum_{j \neq j} u_j, \tag{5.18}$$

near the boundary, with the same assumptions on f_i as before. Then we consider its linearization

$$\left(\frac{\partial}{\partial t} - \Delta\right) v_i = f_i'(u_{i,\kappa}) v_i - \kappa v_i \sum_{j \neq j} u_j - \kappa u_i \sum_{j \neq i} v_j, \tag{5.19}$$

with $v_i = 0$ on $\partial\Omega \times \mathbb{R}$.

First, we want to know, for the equation (5.18), if the solutions $u_{i,\kappa}$ converge to u_i as $\kappa \to +\infty$, such that in $Q_R \cap (\Omega \times \mathbb{R})$, $u_1 > 0$, then what is the behavior of $u_{i,\kappa}$? We have the following result:

Proposition 5.4.1 *Under the above assumptions, for κ large enough,*

$$\sup_{B_R \cap \Omega} \sum_{i>1} u_{i,\kappa}$$

converge to 0 rapidly.

Proof As in Sect. 5.1, we can assume we are in $Q_1 \cap (\Omega \times \mathbb{R})$, with the Lipschitz constant of f_i smaller than a small constant $\theta > 0$, and for a small $h_0 > 0$

$$\sup_{Q_1 \cap (\Omega \times \mathbb{R})} |u_{1,\kappa} - u_1| \leq h_0.$$

Define the iteration (starting with $R_0 = 1$)

$$\begin{cases} h_{l+1} = h_l^2, \\[2mm] R_l = R_{l-1} - h_l^{\frac{1}{2}}. \end{cases}$$

In each $Q_{R_l} \cap (\Omega \times \mathbb{R})$, we can decompose $u_{1,\kappa} := v_l + w_l$, such that

$$
\begin{cases}
\left(\dfrac{\partial}{\partial t} - \Delta \right) v_l = f_1(v_l), & \text{in } Q_{R_l} \cap (\Omega \times \mathbb{R}) \\
v_l = u_{1,\kappa}, & \text{on } \partial_p \big(Q_{R_l} \cap (\Omega \times \mathbb{R}) \big).
\end{cases}
$$

As in Sect. 5.1, such v_l exists and is unique. Moreover, $v_l > 0$ in $Q_{R_l} \cap (\Omega \times \mathbb{R})$.
For $l = 0$, from the stability and interior gradient estimate, we have

$$
\begin{cases}
\sup_{B_1 \cap \Omega} |v_0 - u_1| \le 2h_0, & \text{in } Q_1 \cap (\Omega \times \mathbb{R}) \\
|\nabla v_0 - \nabla u_1| \le h_0^{\frac{1}{2}}, & \text{in } Q_{1-h_0^{\frac{1}{2}}} \cap (\Omega \times \mathbb{R}),
\end{cases}
$$

which implies, in $Q_{1-h_0^{\frac{1}{2}}} \cap (\Omega \times \mathbb{R})$ (here the distance is in the space direction, that is, in a time slice)

$$
v_0(x) \ge C \, dist(x, \partial \Omega \times \mathbb{R}).
$$

Then we can proceed as before, to divide $Q_{1-h_0^{\frac{1}{2}}} \cap (\Omega \times \mathbb{R})$ into two domains: where $v_l \ge Ch_0$ (then $u_{1,\kappa} \ge Ch_0$), and where $v_l \le Ch_0$, which is a narrow domain with width h_0, and we can estimate the decay of $\sum_{j>1} u_{j,\kappa}$ when restricted to $Q_{1-2h_0^{\frac{1}{2}}} \cap (\Omega \times \mathbb{R})$. The detail is exactly as before and we will not repeat it here. $\qquad \square$

Proposition 5.4.2 *Under the assumptions of the previous proposition, and if in $Q_R \cap (\Omega \times \mathbb{R})$, $v_{i,\kappa}$, are solutions of (5.19), satisfying:*

$$
\sum_i \int_{Q_R \cap (\Omega \times \mathbb{R})} |v_{i,\kappa}|^q
$$

are uniformly bounded for some $q > 3n$, then in $Q_{\frac{R}{4}} \cap (\Omega \times \mathbb{R})$

$$
v_{j,\kappa} \to 0 \text{ rapidly}, \quad \text{for } j \ne 1.
$$

Moreover

$$
\left(\frac{\partial}{\partial t} - \Delta \right) v_{1,\kappa} = f_1'(u_{1,\kappa}) v_{1,\kappa} + \varepsilon(\kappa),
$$

where $\lim_{\kappa \to +\infty} \varepsilon(\kappa) = 0$ rapidly.

Proof First, for $i > 1$

$$\left(\frac{\partial}{\partial t} - \Delta\right) v_{i,\kappa}^2 \leq f_i'(u_{i,\kappa}) v_{i,\kappa}^2 - \kappa v_{i,\kappa}^2 \sum_{j \neq i} u_j - \kappa u_i \sum_{j \neq i} v_{i,\kappa} v_{j,\kappa}$$

$$\leq f_i'(u_{i,\kappa}) v_{i,\kappa}^2 - \varepsilon(\kappa) \sum_{j \neq j} v_{i,\kappa} v_{j,\kappa}. \tag{5.20}$$

By noting the boundary condition for $v_{i,\kappa}$, standard parabolic estimate (c.f. [29]) gives, for $i > 1$

$$\sup_{Q_{\frac{R}{2}} \cap (\Omega \times \mathbb{R})} |v_{i,\kappa}| \leq C,$$

for some constant C.

Next, because $u_{1,\kappa}(x) \geq C \, dist(x, \partial\Omega)$, $\forall h > 0$, $\forall x \in \{x : dist(x, \partial\Omega) \geq 2Ch\}$, there exists a cylinder $Q_h(x)$, where $u_{1,\kappa}(x) \geq h$, so by Lemma 5.2.1 and the first inequality in (5.20), we get

$$v_{i,\kappa}^2(x) \leq e^{-\kappa h^3}.$$

While for $x \in \{x : dist(x, \partial\Omega) \leq 2Ch\}$, noting that this is a narrow domain in the sense before, using the second inequality in (5.20), we have a decay from $Q_{\frac{R}{2}-kh} \cap (\Omega \times \mathbb{R})$ to $Q_{\frac{R}{2}-(k+1)h} \cap (\Omega \times \mathbb{R})$, $k = 0, 1, 2, \ldots, [\frac{R}{4h}]$, so we get

$$\sup_{Q_{\frac{R}{4}} \cap \{x:0<dist(x,\partial\Omega)\leq 2Ch\}} v_{i,\kappa}^2(x) \leq 2^{-Rh^{-1}} \sup_{Q_{\frac{R}{2}} \cap \{x:0<dist(x,\partial\Omega)\leq 2Ch\}} v_{i,\kappa}^2$$

$$+ \sup_{Q_{\frac{R}{2}} \cap \{x:dist(x,\partial\Omega)\geq 2Ch\}} v_{i,\kappa}^2.$$

If we choose $h = \kappa^{-\frac{1}{4}}$, we then get the first claim in the proposition. Then the second claim can be easily seen. $\qquad\square$

Remark 5.4.3 In 1 dimension and the elliptic case, the above iteration does not work. We need to use the modification introduced in Remark 5.2.5. Note that we have boundary value 0 in the above proposition.

Chapter 6
Asymptotics in Strong Competition

Abstract This chapter is devoted to the study of the asymptotic behavior of the following system when κ is sufficiently large.

$$\begin{cases} \dfrac{\partial u_i}{\partial t} - \dfrac{\partial^2 u_i}{\partial x^2} = a_i u_i - u_i^2 - \kappa u_i \sum_{j \neq i} b_{ij} u_j & \text{in } [0,1] \times (0,+\infty), \\ u_i = 0 & \text{on } \{0\} \times (0,+\infty) \cup \{1\} \times (0,+\infty), \\ u_i = \phi_i & \text{on } [0,1] \times \{0\}. \end{cases}$$
(6.1)

Here $i = 1, 2, \ldots, M$, ϕ_i are given Lipschitz continuous functions on $[0,1]$ such that

$$\phi_i \geq 0, \quad \text{and if} \quad i \neq j, \ \phi_i \phi_j \equiv 0.$$

For simplicity, we assume the coefficients $b_{ij} = 1$, $\forall i \neq j$. Without this assumption, our proof is still valid with minor changes, due to the special property of dimension 1. In the following, we will point out this whenever necessary. From the discussions in Chap. 4, we know the singular limit of (6.1) as $\kappa \to +\infty$ is a gradient flow (that is, satisfying an energy inequality), with its solution converging to the stationary state as $t \to +\infty$. So the natural question arises: does (6.1) for κ large also behave like a gradient system? In this chapter, we will show that with a few assumptions, the answer is yes: for κ large the dynamics of (6.1) is simple. In the first section, we list some non-degeneracy conditions needed in our proof and give the main result of this chapter. The next section is devoted to a coarse convergent property of solutions to (6.1). Finally, in the last section we use the non-degeneracy conditions to finish the proof of the convergence.

6.1 Non-degeneracy Conditions and the Main Result

We need impose some assumptions on the stationary equation of (4.1), exactly as in [22]. From Sect. 4.1, we know any solution of (4.1) can be arranged properly so that, if we define

$$u := u_1 - u_2 + \cdots,$$

K. Wang, *Free Boundary Problems and Asymptotic Behavior of Singularly Perturbed Partial Differential Equations*, Springer Theses, DOI 10.1007/978-3-642-33696-6_6, © Springer-Verlag Berlin Heidelberg 2013

then u satisfies

$$-\frac{d^2u}{dx^2} = f(x, u) \quad \text{on } [0, 1], \quad u(0) = u(1) = 0, \tag{6.2}$$

where

$$f(x, u) = a_i u - (-1)^{i+1} u^2, \quad \text{for } x \in \{x \in [0, 1] | u_i(x) > 0\}.$$

If $\exists b_{ij} \neq 1$, in the above definition of u, u has the form $u_1 - c_1 u_2 + c_2 u_2 - \cdots$, where $c_1, c_2, \ldots$ are constants depending only on b_{ij}. The following statement and proof can be changed accordingly.

We first give a result about non-degeneracy of solutions of (6.2) (or equivalently, (4.1)).[1]

Proposition 6.1.1 *Any solution u of* (6.2) *is non-degenerate, that is, any solution v of*

$$-v'' = g(x, u)v \quad \text{a.e. on } [0, 1], \quad v(0) = v(1) = 0 \tag{6.3}$$

is identically zero. Here

$$g(x, u) = f'(x, u) = a_i - 2(-1)^{i+1} u, \quad \text{for } x \in \{x \in [0, 1] | u_i(x) > 0\}.$$

Proof In fact, we know that $f(x, u)$ is differentiable in u except at $u = 0$, and $f'_u(x, u) < \frac{f(x,u)}{u}, u \neq 0, f(x, 0) = 0$. (Note that non-trivial solutions of (6.2) only vanish on a set of measure zero so that the linearization (6.3) makes sense.)

If z is a solution of (6.2) with k interior zeroes, then z' is a solution of the linearized equation (6.3), and it must have at least $k + 1$ zeroes in $(0, 1)$, one between each two zeroes of z in $[0, 1]$.

Suppose that v satisfies (6.3). By the Sturm comparison theorem (comparing with z'), v has at least k zeroes in $(0, 1)$ (one between any two zeroes of z').

Now

$$-z'' - \frac{f(x, z)}{z} z = 0 \quad \text{a.e. on } [0, 1], \quad z(0) = z(1) = 0$$

and z has k interior zeroes. Hence, 0 must be the $(k + 1)$th eigenvalue of the eigenvalue problem

$$-h'' - \frac{f(x, z)}{z} h = \lambda h \quad \text{a.e. on } [0, 1], \quad h(0) = h(1) = 0.$$

---

[1] I would like to thank Prof. E.N. Dancer for pointing out this fact to me.

Since $f'_u(x, u) < \frac{f(x,u)}{u}$ as $u \neq 0$ a.e. on $(0, 1)$, it follows by eigenvalue comparison that the $(k + 1)$th eigenvalue of

$$-v'' - g(x, u)v = \lambda v \quad \text{a.e. on } [0, 1], \quad v(0) = v(1) = 0$$

is positive. Thus if 0 is an eigenvalue, it must be the lth eigenvalue where $l < k + 1$. Hence the corresponding eigenfunction has less than $l - 1$ interior zeroes, so less than k interior zeroes. This contradicts the first part of the proof and hence 0 is not an eigenvalue, i.e., (6.3) has no non-trivial solutions. □

Next, we propose some non-degeneracy conditions.

1. The system of differential inequalities

$$\begin{cases} -\dfrac{d^2 u_i}{dx^2} \leq a_i u_i & \text{in } [0, 1], \\ -\dfrac{d^2}{dx^2}\left(u_i - \sum_{j \neq i} u_j\right) \geq a_i u_i - \sum_{j \neq i} a_j u_j & \text{in } [0, 1], \\ u_i(0) = u_i(1) = 0, \\ u_i \geq 0, u_i u_j = 0 & \text{for } i \neq j, \text{ in } [0, 1], \ (1 \leq i, j \leq M) \end{cases}$$

(6.4)

has no nontrivial solutions.

(Assume (u_i) is a nontrivial solution (we still consider a function on disjoint supports as two distinct functions), similarly to Sect. 4.1, there exist α_m ($1 \leq m \leq k$ for some positive integer k) such that

$$0 = \alpha_0 < \alpha_1 < \alpha_2 < \cdots < \alpha_k = 1,$$

and there exist constants $c_m > 0$, $d_m = \frac{-\alpha_{m-1}}{\alpha_m - \alpha_{m-1}}\pi$ ($1 \leq m \leq k$), such that

$$u_{i_m}(x) = \begin{cases} c_m \sin(\frac{\pi}{\alpha_m - \alpha_{m-1}}x + d_m), & \text{if } x \in (\alpha_{m-1}, \alpha_m), \\ 0, & \text{otherwise}; \end{cases}$$

and if $j \neq i_m$ for all m, $u_j \equiv 0$. Moreover, $\forall m$, $c_{m+1}\sqrt{a_{i_{m+1}}} = c_m\sqrt{a_{i_m}}$ and $\alpha_m - \alpha_{m-1} = \frac{\pi}{\sqrt{a_{i_m}}}$. This last condition implies

$$\sum_{m=1}^{k} \frac{\pi}{\sqrt{a_{i_m}}} = 1.$$

This is very restrictive. So we can assume, $\exists\{a_i, i = 1, 2, \ldots, M\}$ such that there is only trivial solution of (6.4).)

2. Non-constant (in time) bounded positive solutions of the following system ($i = 1, 2, \ldots, M$)

$$
\begin{cases}
\dfrac{\partial u_1}{\partial t} - \dfrac{\partial^2 u_1}{\partial x^2} = a_1 u_1 - u_1 \displaystyle\sum_{j \neq 1} u_j & \text{in } [0, 1] \times (-\infty, +\infty), \\[3mm]
\dfrac{\partial u_2}{\partial t} - \dfrac{\partial^2 u_2}{\partial x^2} = a_2 u_2 - u_2 \displaystyle\sum_{j \neq 2} u_j & \text{in } [0, 1] \times (-\infty, +\infty), \\[2mm]
\vdots & \\[2mm]
\dfrac{\partial u_M}{\partial t} - \dfrac{\partial^2 u_M}{\partial x^2} = a_M u_M - u_M \displaystyle\sum_{j \neq M} u_j & \text{in } [0, 1] \times (-\infty, +\infty), \\[3mm]
u_i(0, t) = u_i(1, t) = 0 & \text{for } t \in (-\infty, +\infty),
\end{cases}
\tag{6.5}
$$

approach distinct stationary solutions as $t \to \pm\infty$. The positive stationary solutions are hyperbolic, and there are no circuits of heteroclinic positive orbits.

(This condition has been studied in [22], where they construct an example (with two equations) satisfying this condition.)

3. $a_i > \pi^2$ ($\forall i = 1, 2, \ldots, M$, π^2 is the first eigenvalue of $-w'' = \lambda w, w(0) = w(1) = 0$)

(Assumption 3 implies that the equation ($\forall i = 1, 2, \ldots, M$)

$$
\begin{cases}
-\dfrac{d^2 u}{dx^2} = a_i u & \text{in } (0, 1), \\[3mm]
u(0) = u(1) = 0,
\end{cases}
\tag{6.6}
$$

has no positive solutions.)

Now we state our main result.

Theorem 6.1.2 *Under the above Assumptions 1–3, for κ large, any solution of (6.1) converges to a stationary point as $t \to +\infty$.*

6.2 A Coarse Convergence Result

In this section, we will establish a convergence result, which says for κ large, any solution of (6.1), after sufficiently large time, will stay near a solution of (4.1) (a stationary solution of the singular limit problem (4.6)).

First, we need to establish a theorem analogue to the Theorem 2 in [22]. However, their proof cannot be applied here directly, so we need some improvements.

Theorem 6.2.1 $\forall \varepsilon > 0$, *there exists* $\kappa(\varepsilon)$ *such that if* $\kappa > \kappa(\varepsilon)$, *then there is a stationary solution* w *of* (4.6) *such that*

$$\left| u_i(x, t) - w_i(x) \right| \leq \varepsilon$$

for all $x \in [0, 1]$ *if* t *is large*.

Proof Take the space $X := C^\alpha([0, 1], \mathbb{R}^M)$ for some fixed $\alpha \in (0, 1)$. For the solution u_κ, define

$$A_\kappa := \bigcup_t \{ u_\kappa(t), \ t \in [0, +\infty) \}.$$

Because $u_\kappa(t)$ are uniformly bounded in $C^\beta([0, 1], \mathbb{R}^M)$ for any $\beta \in (0, 1)$ with bounds independent of κ and t (see, [2] for example, this can also be proved by the blow up method in [11]), we know that $\bigcup_\kappa A_\kappa$ is a pre-compact set of X. So we can define the limit set

$$A := \{ u \in X : \exists v_\kappa \in A_\kappa, \text{ such that } v_\kappa \to u \text{ in } X \}.$$

Claim 1 A is a compact, closed set composed by the connecting orbits of the limit equation (4.6) with corresponding stationary points.

Here we add a remark: because we are in the situation of initial value problems, the "connecting orbit" may connect the initial value to a stationary point of (4.6). However, this only causes minor changes in the following proof, because the local compactness and uniform convergence still hold. The discussion below can be modified slightly to deal with this case and we will omit it.

The compactness is from the fact that $\| v_\kappa \|_{C^\beta} \leq C(\beta)$ and $v_\kappa \to u$ in $C[0, 1]$, which implies $\| u \|_{C^\beta} \leq C(\beta)$, $\forall u \in A$.

The second claim that A is invariant can be shown as follows: assume $u_\kappa = u_\kappa(t_\kappa) \to u$; first we have for $i \neq j$

$$\lim_{\kappa \to \infty} \sup_{[0,1] \times [0, +\infty)} u_{i,\kappa} u_{j,\kappa} = 0,$$

so $u_i u_j \equiv 0$ for $i \neq j$; next define

$$v_\kappa(t) = u_\kappa(t_\kappa + t),$$

then standard method shows that v_κ locally converges (in the sense of [5]) to a solution $v(t)$ of (4.6) with $v(0) = u$.

If t_κ is bounded, then v is defined on $[-T_0, +\infty)$ for some $T_0 > 0$ (noting that for any $T > 0$, $u_\kappa(t)$ converge uniformly on $[0, T]$). Here, we must have $v(-T_0) = \varphi$, the initial value.

If $t_\kappa \to +\infty$, then v is defined on $(-\infty, +\infty)$. We need to prove that v is a connecting orbit. In view of Proposition 4.2.11, we just need to prove that

$$\lim_{t \to -\infty} E(v, t) < +\infty,$$

where the energy E is defined before Proposition 4.2.9. This can be guaranteed by the uniform Lipschitz continuity of $u_\kappa(t)$ for all $\kappa > 0$ and $t > 0$, which implies the existence of a constant $C > 0$, such that for any $t \in (-\infty, +\infty)$, the Lipschitz constant of $v(t)$ is bounded by C. However, if we are in the case $b_{ij} \neq b_{ji}$, we do not know whether the uniform Lipschitz continuity is true, we need to proceed as follows.

In view of the monotonicity of $E(t)$ in the limit case, we just need to prove that

$$\lim_{t \to -\infty} \int_t^{t+1} E(v, \tau)\, d\tau < +\infty.$$

This can be seen by taking the limit in $\int_t^{t+1} E(u_{\kappa_n}, \tau)\, d\tau$. First we know, as a measure $\kappa_n u_{i,\kappa_n} \sum_{j \neq i} u_{j,\kappa_n}$ converge weakly to a Radon measure ν_i, which is supported on $\partial\{v_i > 0\}$. With the convergence of u_{κ_n} to v in $C([T_1, T_2] \times [0, 1])$ ($T_1 \leq T_2$ given), we get

$$\lim_{n \to +\infty} \int_{T_1}^{T_2} \int_{[0,1]} \kappa_n u_{i,\kappa_n}^2 \sum_{j \neq i} u_{j,\kappa_n}\, dx\, dt$$

$$= \lim_{n \to +\infty} \int_{T_1}^{T_2} \int_{[0,1]} \kappa_n v_i u_{i,\kappa_n} \sum_{j \neq i} u_{j,\kappa_n}\, dx\, dt$$

$$= \int_{T_1}^{T_2} \int_{[0,1]} v_i\, d\nu_i$$

$$= 0.$$

By multiplying the equation with u_{i,κ_n}, we get

$$\int_{[0,1]} \frac{1}{2} u_{i,\kappa_n}^2(T_2) - \int_{[0,1]} \frac{1}{2} u_{i,\kappa_n}^2(T_1) + \int_{T_1}^{T_2} \int_{[0,1]} |\nabla u_{i,\kappa_n}|^2$$

$$= \int_{T_1}^{T_2} \int_{[0,1]} a_i u_{i,\kappa_n}^2 - \frac{1}{3} u_{i,\kappa_n}^3 - \kappa_n u_{i,\kappa_n}^2 \sum_{j \neq i} u_{j,\kappa_n}. \tag{6.7}$$

By taking the limit, and noting that other terms converge, we get

$$\lim_{n \to +\infty} \int_{T_1}^{T_2} \int_{[0,1]} |\nabla u_{i,\kappa_n}|^2 = \int_{T_1}^{T_2} \int_{[0,1]} |\nabla v_i|^2.$$

Moreover, from (6.7), we see that the above quantities are uniformly bounded depending only on $T_2 - T_1$ and $\sup \sum_i u_i$. Thus,

$$\int_t^{t+1} \int_{[0,1]} \sum_i \int_{[0,1]} \frac{1}{2} \left| \frac{\partial v_i}{\partial x} \right|^2 - \frac{1}{2} a_i v_i^2 + \frac{1}{3} v_i^3$$

are uniformly bounded independent of t. This finishes the verification of Claim 1.

Now assume that all the solutions of (4.1) are $\{w_0, w_1, \ldots\}$ (by Theorem 4.1.4, this is a finite set) with $w_0 = 0$ (that is, all of the components are identically zero). In the space X, take small open neighborhoods V_i for each w_i, and take a neighborhood U of A. We know that for κ large, $A_\kappa \subset U$ (by the compactness). We can also take an open neighborhood V_{-1} of the initial value φ, and denote φ by w_{-1}.

Claim 2 There exist two universal constants $T > T'$ (depending on our choice of the open neighborhoods V_i only), such that for any connecting orbit of (4.6), its time lying outside $\bigcup_i V_i$ is smaller than T and greater than T'.

This can be proved by a compactness argument (note that all of the connecting orbits form a compact set, and the stationary points are finite).

Claim 3 $\forall \varepsilon > 0$, $\exists \kappa_0$, such that if $\kappa > \kappa_0$, then there exist some connecting orbits $v_1(t), v_2(t), \ldots$ of (4.6), with either $u_\kappa(t) \in \bigcup_i V_i$ or for some t_κ

$$\sup_{[-\frac{T}{2}, \frac{T}{2}]} \left\| u_\kappa(t) - v_i(t - t_\kappa) \right\| \le \varepsilon.$$

Here T is the constant in Claim 2. This roughly says that outside $\bigcup_i V_i$, $u_\kappa(t)$ can not be rotating too much around the manifold of connecting orbits: it almost goes down with a connecting orbit directly. Later we will show that the number of these connecting orbits are finite, but at this stage we can not exclude the possibility that there is an infinite number of them. However, at least we can choose a countable number of these connecting orbits (each occupy a period of time T, where $T > 0$ is a fixed constant).

The proof of Claim 3 is easy: if $u_\kappa(t_\kappa) \to v \in A$ with v not in $\bigcup_i V_i$, then from compactness, on $[-\frac{T}{2}, \frac{T}{2}]$

$$u_\kappa(t + t_\kappa) \to v(t),$$

where $v(0) = v$.

Claim 4 $\exists \delta > 0$, depending only on our choice of V_i, such that for any connecting orbit v, if for $t \in [-\frac{T'}{2}, \frac{T'}{2}]$, $v(t)$ lies outside $\bigcup_i V_i$, then we have

$$E\left(v, \frac{T'}{2}\right) \le E\left(v, -\frac{T'}{2}\right) - \delta.$$

This can be proven by the compactness.

Claim 5 $\exists \kappa_0$, such that for $\forall \kappa \ge \kappa_0$, if for $t \in [-\frac{T'}{2}, \frac{T'}{2}]$, $u_\kappa(t)$ lies outside $\bigcup_i V_i$, then there exists $h > 0$ independent of κ, such that

$$\int_{\frac{T'}{2}}^{\frac{T'}{2}+h} E(u_\kappa, t)\, dt \le \int_{-\frac{T'}{2}}^{-\frac{T'}{2}+h} E(u_\kappa, t)\, dt - \frac{\delta}{2}.$$

Assume this is wrong, then there exists a sequence $\kappa_n \to +\infty$, such that u_{κ_n} is the solution of (6.1) with the fixed initial value ϕ, and $u_{\kappa_n}([t_{\kappa_n} - \frac{T'}{2}, t_{\kappa_n} + \frac{T'}{2}])$ lie outside $\bigcup_i V_i$, but

$$\int_{t_{\kappa_n} + \frac{T'}{2}}^{t_{\kappa_n} + \frac{T'}{2} + h} E(u_\kappa, t) \, dt \geq \int_{t_{\kappa_n} - \frac{T'}{2}}^{t_{\kappa_n} - \frac{T'}{2} + h} E(u_\kappa, t) \, dt - \frac{\delta}{2}. \tag{6.8}$$

If we assume the uniform Lipschitz estimate, then for all $t \in [-\frac{T'}{2}, \frac{T'}{2}]$ and n, $u_{\kappa_n}(t_{\kappa_n} + t)$ are uniformly Lipschitz continuous. So after taking a subsequence, it converges to a solution $v(t)$ of (4.6), which is defined on $[-\frac{T'}{2}, \frac{T'}{2}]$, and $v(t)$ lie outside $\bigcup_i V_i$ ($\bigcup_i V_i$ is an open set), and

$$\int_{\frac{T'}{2}}^{\frac{T'}{2} + h} E(v, t) \, dt \geq \int_{-\frac{T'}{2}}^{-\frac{T'}{2} + h} E(v, t) \, dt - \frac{\delta}{2}. \tag{6.9}$$

This contradicts Claim 4. In fact, in the above in order to guarantee the quantities in (6.8) converge to those in (6.9), we do not need the stronger condition on uniform Lipschitz continuity, see the last part of the proof of Claim 1. Thus, our Claim 5 follows.

Claim 6 If we choose V_i small enough, then $\exists h > 0$, κ_0 large enough, such that for any solution u_κ of (6.1) with $\kappa > \kappa_0$, if $u_\kappa([T_1 - h, T_1])$ and $u_\kappa([T_2, T_2 + h])$ lie in V_i (here $T_2 > T_1$), then

$$\int_{T_2}^{T_2 + h} E(u_\kappa, t) \, dt \leq \int_{T_1 - h}^{T_1} E(u_\kappa, t) \, dt + \frac{\delta}{4}.$$

By (6.7), the gradient terms in the above integral can be transformed into those terms containing $u_{i,\kappa}$ only (no involvement in gradients of u_i). However, $u_{i,\kappa}$ are uniformly Hölder continuous with respect to the parabolic distance, so if we choose V_i and h small enough (depending on δ only), then for any $t \in [0, h]$

$$\sup_\Omega |u_\kappa(T_1 - h + t) - u_\kappa(T_2 + t)| \leq \frac{\delta}{16}.$$

Finally, we note that the last term in (6.7) converge to 0 as $\kappa \to +\infty$ (using compactness, we can prove that this convergence is uniform, that is, $\forall \varepsilon > 0$, $\exists \kappa_0$, such that $\forall \kappa > \kappa_0$, this term is smaller than ε), and Claim 6 follows.

Now we have the following picture. If $u_\kappa(t_0)$ is not in $\bigcup_i V_i$, because it is close to some connecting orbit $v(t)$, from Claim 1, we know after a time $\leq 2T$, u_κ will enter some V_i. Moreover, its energy decays by an amount at least $-\frac{\delta}{2}$.

If $u_\kappa(t_0) \in V_i$, then either for any $t > t_0$, we have $u_\kappa(t) \in V_i$ (staying inside and not getting out), then in view of our arbitrary choice of V_i, we conclude; or there exists a $t_1 > t_0$ such that $u_\kappa(t_1)$ gets outside of V_i. Then after a time T, it will get into some V_i again and its energy decay by an amount at least $-\frac{\delta}{2}$. However, we know the energy $E(u_\kappa, t)$ is bounded, so these procedures can happen at most finitely many times, and after some time it will stay in some V_i and never gets outside of it, that is there exists a $t^* > 0$, such that for any $t > t^*$, we have $u_\kappa(t) \in V_i$ for some i. In view of the arbitrary choice of V_i, we conclude. $\qquad\square$

The above theorem can be used to prove the following corollary.

Corollary 6.2.2 *For κ large, any periodic solution of (6.1) must stay near some stationary point of the singular limit system (4.6) in $C^\alpha[0, 1]$ for any $\alpha \in (0, 1)$.*

Proof We just need to note that for a periodic solution, if for large time t, it stays in an open neighborhood V_i of some w_i (same notations as in the previous theorem), then for all time t, it stays in this open neighborhood. $\qquad\square$

6.3 The Refined Convergence Result

Now we study the structure of orbits for κ large, near a nontrivial stationary point of the limit equation. This will give the full convergence of solutions to (6.1).

Theorem 6.3.1 *Assume that w is a non-trivial stationary solution of (4.6). Then there exists ε, $\kappa_0 > 0$ such that if $u_{i,\kappa}$ is a solution of (6.1) for $\kappa > \kappa_0$ with $\|u_{i,\kappa} - w_i\|_\infty \leq \varepsilon$ for all large t, then $u_{i,\kappa}(t) \to v_i$ uniformly on $[0, 1]$ as $t \to +\infty$, where v_i is a nonnegative nontrivial stationary solution of (6.1) with v_i near w_i (in $L^\infty([0, 1])$).*

Proof By using omega limit sets, we see that it suffices to prove that the only solutions u_κ of (6.1) defined for all t and satisfying $\|u_{i,\kappa} - w_i\|_\infty \leq \varepsilon$ for all t are the constant solutions (stationary in time). Suppose this is false.

We first consider fixed κ. If $u_{i,\kappa}$ is a non-stationary solution of (6.1) which is bounded for all t; standard local parabolic estimates imply that $\frac{\partial u_{i,\kappa}}{\partial t}$ are uniformly bounded (and at least one is nontrivial).

By differentiating the equation in time t, we get on $[0, 1] \times \mathbb{R}$

$$\left(\frac{\partial}{\partial t} - \frac{\partial^2}{\partial x^2}\right)\frac{\partial u_{i,\kappa}}{\partial t} = (a_i - 2u_{i,\kappa})\frac{\partial u_{i,\kappa}}{\partial t} - \kappa\frac{\partial u_{i,\kappa}}{\partial t}\sum_{j\neq i}u_{j,\kappa} - \kappa u_{i,\kappa}\sum_{j\neq i}\frac{\partial u_{j,\kappa}}{\partial t}.$$

We can rescale $\frac{\partial u_{i,\kappa}}{\partial t}$ to $v_{i,\kappa}$ so that (after a translation in time)

$$\sup_t \sum_i \int_t^{t+1}\int_0^1 v_{i,\kappa}^{10} = 1, \qquad (6.10)$$

$$\sum_i \int_0^1\int_0^1 v_{i,\kappa}^{10} \geq \frac{1}{2}. \qquad (6.11)$$

With this integral bound, by the proof of Proposition 5.3.3 (in the interior) and Proposition 5.4.2 (near the boundary), we get

$$\sup_{[0,1]\times(-\infty,+\infty)} \sum_i v_{i,\kappa}^2 \le C, \tag{6.12}$$

for some constant C independent of κ.

Now let $\kappa \to +\infty$, we can prove in some weak sense specified below

$$u_{i,\kappa}(x,t) \to u_i(x,t),$$

$$v_{i,\kappa}(x,t) \to v_i(x,t),$$

where

1. $u_i(x,t) = w_i(x)$, this is because: first we have for $\forall t \in (-\infty,+\infty)$

$$\left\| u_i(t) - w_i \right\|_\infty \le \varepsilon;$$

then by Proposition 4.2.11 and the fact that w is isolated, we have

$$\lim_{t\to\pm\infty} u_i(x,t) = w_i(x);$$

now we can use the energy identity (Proposition 4.2.9) to conclude that $\frac{\partial u_i}{\partial t} \equiv 0$.

2. $v_i(x,t) \equiv 0$ outside $\{w_i > 0\}$. This is an easy consequence of Lemma 5.2.1, Proposition 5.3.3 (the case near the free boundary) and Proposition 5.4.2 (dealing with the boundary point).

3. In $\{w_i > 0\} \times (-\infty,+\infty)$

$$\left(\frac{\partial}{\partial t} - \frac{\partial^2}{\partial x^2}\right) v_i = a_i v_i - 2w_i v_i.$$

This is because in the equation of $v_{i,\kappa}$, for $j \ne i$, $u_{j,\kappa}$ and $v_{j,\kappa}$ converge to 0 rapidly in any compact subset of this domain, thus we can take the limit.

4. Near the regular part of $\partial\{w_i > 0\} \cap \partial\{w_j > 0\}$ (here all of the free boundaries are regular: they are lines), we have

$$\left(\frac{\partial}{\partial t} - \frac{\partial^2}{\partial x^2}\right)(v_i - v_j) = a_i v_i - 2w_i v_i - a_j v_j + 2w_j v_j.$$

This can be proved by Corollary 5.3.2.

5.

$$\sum_i \int_0^1 \int_0^1 v_i^{10} \ge \frac{1}{4}. \tag{6.13}$$

This can be proved by the combination of (6.10) and (6.12), which implies for a small $\delta > 0$,

$$\sum_i \int\int_{[0,1]\times[0,1]\setminus\{\sum_i w_i \le \delta\}} v_{i,\kappa}^{10} \ge \frac{1}{4}.$$

Then because in $[0, 1] \times [0, 1] \setminus \{\sum_i w_i \leq \delta\}$, $v_{i,\kappa}$ converge to v_i uniformly, we can take the limit to get

$$\sum_i \int \int_{[0,1] \times [0,1] \setminus \{\sum_i w_i \leq \delta\}} v_i^{10} \geq \frac{1}{4}.$$

In our situation, we can rearrange v_i so that if we define

$$v := v_1 - v_2 + \cdots,$$

then v satisfies

$$\left(\frac{\partial}{\partial t} - \frac{\partial^2}{\partial x^2}\right) v = g(x, w)v,$$

with $g(x, w) := (-1)^{i+1}(a_i - 2w_i)$, for $x \in \{x \in [0, 1] | w_i(x) > 0\}$. Using the expansion with eigenfunctions of the operator $\frac{d^2}{dx^2} + g(x, w)$, from the nondegeneracy of (6.2) and the boundedness of v, we can show

$$v(x, t) \equiv 0,$$

which contradicts (6.13). $\qquad\square$

Remark 6.3.2 In the proof of the uniform boundedness, (6.12), from the bound on the L^{10} norm, (6.10), we can also use the Kato inequality, as in [22], that is, by the equations satisfied by $v_{i,\kappa}$, we get

$$\left(\frac{\partial}{\partial t} - \frac{\partial^2}{\partial x^2}\right) |v_{i,\kappa}| \leq (a_i - 2u_{i,\kappa})|v_{i,\kappa}| - \kappa|v_{i,\kappa}| \sum_{j \neq i} u_{j,\kappa} + \kappa u_{i,\kappa} \sum_{j \neq i} |v_{j,\kappa}|.$$

Summing over i, the last two terms are canceled and we get

$$\left(\frac{\partial}{\partial t} - \frac{\partial^2}{\partial x^2}\right) \sum_i |v_{i,\kappa}| \leq \sum_i (a_i - 2u_{i,\kappa})|v_{i,\kappa}|.$$

Then standard parabolic estimates give our bound on $\sup |v_{i,\kappa}|$.

Unfortunately, it seems impossible to extend this method to the setting:

$$\left(\frac{\partial}{\partial t} - \frac{\partial^2}{\partial x^2}\right) v_{i,\kappa} = (a_i - 2u_{i,\kappa})v_{i,\kappa} - \kappa v_{i,\kappa} \sum_{j \neq i} b_{ij} u_{j,\kappa} + \kappa u_{i,\kappa} \sum_{j \neq i} b_{ij} v_{j,\kappa},$$

where b_{ij} need not equal b_{ji}. On the other hand, our proof can be carried out in this case.

Corollary 6.3.3 *Any nontrivial periodic solution of* (6.1) *for κ large must stay near 0. In particular, its sup norm is small.*

Proof First, the method of Theorem 6.2.1 gives that for any solution u_κ of (6.1), which is defined on $(-\infty, +\infty)$, there exists a $T > 0$ and a stationary point w of (4.6), such that for any $t > T$, $u_\kappa(t)$ lies near w. In our situation of periodic solutions, it implies that for all $t \in (-\infty, +\infty)$, $u_\kappa(t)$ lies near w. Now Theorem 6.3.1 says that if $w \neq 0$, u_κ must be a stationary solution.   □

Finally, we consider the case for those solutions staying near 0.

Theorem 6.3.4 *With the Assumptions* 1, 2, 3, *there exist* ε, $\kappa_0 > 0$ *such that if* $u_{i,\kappa}$ *is a solution of* (6.1) *for* $\kappa > \kappa_0$ *with* $\|u_{i,\kappa}\|_\infty \leq \varepsilon$ *for all large* t, *then* $u_{i,\kappa}(t)$ *converges uniformly on* $[0, 1]$ *as* $t \to +\infty$.

Proof As in Theorem 6.3.1, we only need to consider solutions defined on all time. We can assume (by translation in time)

$$\sup_t \sum_i \int_t^{t+1} \int_0^1 u_{i,\kappa}^2 = \varepsilon(\kappa), \tag{6.14}$$

$$\sum_i \int_0^1 \int_0^1 u_{i,\kappa}^2 \geq \frac{1}{2} \varepsilon(\kappa). \tag{6.15}$$

We define $v_{i,\kappa} = \frac{1}{\varepsilon(\kappa)} u_{i,\kappa}$, which satisfies

$$\frac{\partial v_{i,\kappa}}{\partial t} - \frac{\partial^2 v_{i,\kappa}}{\partial x^2} = a_i v_{i,\kappa} - \varepsilon(\kappa) v_{i,\kappa}^2 - \kappa \varepsilon(\kappa) v_{i,\kappa} \sum_{j \neq i} v_{j,\kappa} \quad \text{on } [0, 1] \times \mathbb{R}.$$

Taking the limit, we get three cases:

1. $\kappa \varepsilon(\kappa) \to 0$;
2. $\kappa \varepsilon(\kappa) \to +\infty$;
3. $\kappa \varepsilon(\kappa) \to \lambda$ for some positive constant λ.

We can prove in all of these cases $v_{i,\kappa}$ are uniformly C^α continuous with respect to the parabolic distance (the cases 1 and 3 are easy and we in fact have higher uniform regularity, and case 2 can be proved using the same method of [2] or simply by the blow up method in [11]). So we can say as $\kappa \to +\infty$, $v_{i,\kappa} \to v_i$ locally uniformly and they satisfy the limit equation weakly. Moreover by taking the limit in (6.14), we have

$$\sup_t \sum_i \int_t^{t+1} \int_0^1 v_i^2 = 1,$$

$$\sum_i \int_0^1 \int_0^1 v_i^2 \geq \frac{1}{2}.$$

This implies global Lipschitz continuity of v_i with respect to the parabolic distance.
   We study these cases separately:

Case 1 Here we have the limit equation

$$\frac{\partial v_i}{\partial t} - \frac{\partial^2 v_i}{\partial x^2} = a_i v_i \quad \text{on } [0, 1] \times \mathbb{R}.$$

We can easily prove, using the energy identity, that as $t \to \pm\infty$, $v_i(x, t)$ converge to the nontrivial stationary solutions. Then by the uniqueness of the solution of (6.6) (because they are linear ODEs) and the energy identity again, we know $v_i(x, t) = v_i(x)$. This is a nontrivial solution of (6.6), and we get a contradiction to Assumption 3.

Case 2 Here the limit equation is

$$\begin{cases} \dfrac{\partial v_i}{\partial t} - \dfrac{\partial^2 v_i}{\partial x^2} \leq a_i v_i & \text{on } [0, 1] \times \mathbb{R}, \\[2mm] \left(\dfrac{\partial}{\partial t} - \dfrac{\partial^2}{\partial x^2} \right) \left(v_i - \displaystyle\sum_{j \neq i} v_j \right) \geq a_i v_i - \displaystyle\sum_{j \neq i} a_j v_j & \text{on } [0, 1] \times \mathbb{R}, \end{cases}$$

where v_i have disjoint support.

Here we can prove an energy identity exactly as in Proposition 4.2.9, and then the same method in the Case 1 gives a contradiction to our Assumption 1.

Case 3 Here we have the limit equation

$$\frac{\partial v_i}{\partial t} - \frac{\partial^2 v_i}{\partial x^2} = a_i v_i - \lambda v_i \sum_{j \neq i} v_j \quad \text{on } [0, 1] \times \mathbb{R},$$

for some $\lambda > 0$. By defining $\hat{v}_i = \lambda v_i$, we get

$$\frac{\partial \hat{v}_i}{\partial t} - \frac{\partial^2 \hat{v}_i}{\partial x^2} = a_i \hat{v}_i - \hat{v}_i \sum_{j \neq i} \hat{v}_i \quad \text{on } [0, 1] \times \mathbb{R}.$$

So this case can be treated similarly as in [22] (cf. the last part of Theorem 4 in p. 484) to get a contradiction to Assumption 2. □

Remark 6.3.5 We can pursue the higher dimensional case following this approach. The difficulty in higher dimension is, generally, we do not know if there are finitely many solutions of the elliptic singular system. We also do not know if there are finitely many critical values of the corresponding functional. Note that if we impose nondegenerate conditions as in Sect. 6.1, most part of the analysis in this chapter can be carried out.

Chapter 7
The Limit Equation of a Singularly Perturbed System

Abstract In this chapter we establish the limit equations of the singularly perturbed elliptic and parabolic systems arising in the study of Bose–Einstein condensation. The proof relies on a stationary condition and a monotonicity formula.

7.1 The Elliptic Case

In this section, we consider the limit equation of the following system as $\kappa \to +\infty$:

$$-\Delta u_i = f_i(u_i) - \kappa u_i \sum_{j \neq i} b_{ij} u_j^2, \quad \text{in } B_1(0), \tag{7.1}$$

where $b_{ij} > 0$ are constants which satisfy $b_{ij} = b_{ji}$, $1 \leq i, j \leq M$. $B_1(0)$ is the unit ball in $\mathbb{R}^n$ ($n \geq 1$). Typical model for $f_i(u_i)$ is $f_i(u) = a_i u - u^p$ with constants $a_i > 0$, $p > 1$. We only consider positive solutions, that is, those $u_i \geq 0$ in its domain for all i. We will denote the solution corresponding to κ as $u_\kappa = (u_{1,\kappa}, u_{2,\kappa}, \ldots, u_{M,\kappa})$. Solutions to the above problem are critical points of the following functional:

$$J_\kappa(u) = \int_\Omega \frac{1}{2} \sum_i |\nabla u_i|^2 + \frac{\kappa}{4} \sum_{i \neq j} b_{ij} u_i^2 u_j^2 - \sum_i F_i(u_i), \tag{7.2}$$

where $F_i(u) = \int_0^u f_i(t) \, dt$. As described in the introduction, the expected limit equation of (7.1) as $\kappa \to +\infty$ should be

$$\begin{cases} -\Delta u_i \leq f_i(u_i), & \text{in } B_1(0), \\ -\Delta \left(u_i - \sum_{j \neq i} u_j \right) \geq f_i(u_i) - \sum_{j \neq i} f_j(u_j), & \text{in } B_1(0), \\ u_i \geq 0, & \text{in } B_1(0), \\ u_i u_j = 0, & \text{in } B_1(0). \end{cases} \tag{7.3}$$

In this section, we will verify the above conjecture.

K. Wang, *Free Boundary Problems and Asymptotic Behavior of Singularly Perturbed Partial Differential Equations*, Springer Theses, DOI 10.1007/978-3-642-33696-6_7, © Springer-Verlag Berlin Heidelberg 2013

Theorem 7.1.1 *As $\kappa \to +\infty$, any bounded solution u_κ of (7.1) converges to u, which satisfies the system (7.3).*

Remark 7.1.2 If $f_i(u) = a_i u - u^p$ where $a_i > 0$, $p > 1$ are constants, u_κ are positive solutions of (7.1) with suitable boundary conditions, $u_{i,\kappa}$ are uniformly bounded in κ.

For fixed $\kappa < +\infty$, because the solution u_κ is bounded in $B_1(0)$, standard elliptic estimates show that it is smooth. Since it is the critical point of the functional (7.2), we can consider the following domain variation of u_κ. Take a compactly supported vector field $Y \in C_0^\infty(B_1(0))$; define

$$u_\kappa^s(x) = u_\kappa(x + sY(x)).$$

It is well defined for $|s|$ small and smooth in $B_1(0)$.

Now by the definition of critical points, we have

$$\frac{d}{ds} J_\kappa(u_\kappa^s)\bigg|_{s=0} = 0.$$

From this condition, by a well-known computation (analogues to the derivation of monotonicity formula in harmonic map or Yang-Mills field), we get

$$\int_\Omega \left(\frac{1}{2} \sum_i |\nabla u_{i,\kappa}|^2 + \frac{\kappa}{4} \sum_{i \neq j} u_{i,\kappa}^2 u_{j,\kappa}^2 - \sum_i F_i(u_{i,\kappa}) \right) \mathrm{div}\, Y$$

$$- \sum_i \nabla Y(\nabla u_{i,\kappa}, \nabla u_{i,\kappa}) = 0. \tag{7.4}$$

Written in coordinates, the divergence of a vector field is

$$\mathrm{div}\, Y = \sum_{i=1,\ldots,n} \frac{\partial Y_i}{\partial x_i},$$

and for a function v

$$\nabla Y(\nabla v, \nabla v) = \sum_{i,j=1,\ldots,n} \frac{\partial Y_i}{\partial x_j} \frac{\partial v}{\partial x_i} \frac{\partial v}{\partial x_j}.$$

Note that, with our assumption on f_i, from the above identity we can derive the following famous monotonicity formula (for another approach, see [4]): $\exists$ a constant $C > 0$ independent of κ, such that

$$e^{Cr} r^{2-n} \int_{B_r} \frac{1}{2} \sum_i |\nabla u_{i,\kappa}|^2 + \frac{\kappa}{4} \sum_{i \neq j} u_{i,\kappa}^2 u_{j,\kappa}^2 \tag{7.5}$$

is nondecreasing in r.

We know, as $\kappa \to +\infty$, $u_\kappa \to u$ strongly in $H^1(B_1(0))$ (see [4]), and uniformly in $C^\alpha(B_1(0))$ (see [36]). We also have (see Sect. 2 in [4])

$$\lim_{\kappa \to +\infty} \int_{B_1} \kappa \sum_{i \neq j} u_{i,\kappa}^2 u_{j,\kappa}^2 = 0.$$

For fixed vector field Y, we can take the limit $\kappa \to +\infty$ in (7.4) to get a stationary condition for the limit u. Noting the fact that both div Y and ∇Y are smooth with compact supports, by taking the limit in (7.4) we see that the limit u satisfies

$$\int_\Omega \left(\frac{1}{2} \sum_i |\nabla u_i|^2 - \sum_i F_i(u_i) \right) \operatorname{div} Y - \sum_i \nabla Y (\nabla u_i, \nabla u_i) = 0. \tag{7.6}$$

Thus the monotonicity formula still holds:

$$e^{Cr} r^{2-n} \int_{B_r} \frac{1}{2} \sum_i |\nabla u_i|^2 \tag{7.7}$$

is nondecreasing in r.

From this, the methods of [4] can be applied to derive the following results:

1. All of the u_i are Lipschitz continuous. For a proof, see [36].
2. The Clean Up lemma holds: in a neighborhood of a point $x \in \bigcup_i \partial\{u_i > 0\}$, where

$$\lim_{r \to 0} e^{Cr} \frac{r \int_{B_r(x)} \sum_i |\nabla u_i|^2}{\int_{\partial B_r(x)} \sum_i |u_i|^2} = 1, \tag{7.8}$$

there are only two components of u non-vanishing here.[1] Moreover, the free boundary near this point is $C^{1,\beta}$ smooth for some $\beta \in (0, 1)$. Such points are called regular points of the free boundary.
3. The singular set of the free boundaries has Hausdorff dimension $n - 2$. For a proof, see [42].

Now we have the equation for u_i:

$$-\Delta u_i = f_i(u_i) - |\nabla u_i| H^{n-1} \lfloor_{\partial\{u_i > 0\}}.$$

Here $H^{n-1} \lfloor_{\partial\{u_i > 0\}}$ is the $n - 1$ dimensional Hausdorff measure supported on the regular part of $\partial\{u_i > 0\}$.

In order to prove our main theorem, we need only to establish the following result.

---

[1]In [21], we exclude the possibility to have one component of u, say u_1, vanishing on a locally smooth hypersurface, where u_1 is strictly positive in a deleted neighborhood of this hypersurface, so called multiplicity 1 points on the free boundary. This result is essential to prove that (7.3) is the limit of (7.1).

Proposition 7.1.3 *On the regular part of $\partial\{u_i > 0\} \cap \partial\{u_j > 0\}$*

$$|\nabla u_i| = |\nabla u_j|.$$

Remark 7.1.4 Because the regular part of the free boundary is $C^{1,\beta}$, u_i are $C^{1,\beta}$ continuous up to the regular part of the free boundaries, thus it make sense to say the value of ∇u_i on the regular part of $\partial\{u_i > 0\}$.

Proof Suppose in $B_1(0)$, there are only two components u_1 and u_2 non-vanishing, and the free boundary $\partial\{u_i > 0\} \cap \partial\{u_j > 0\}$ is a $C^{1,\beta}$ hypersurface.

For $\forall Y \in C_0^\infty(B_1(0), \mathbb{R}^n)$, by substituting it into (7.6), we get

$$\int_{B_1(0)} \left(\frac{1}{2} \sum_{i=1,2} |\nabla u_i|^2 - \sum_i F_i(u_i)\right) \operatorname{div} Y - \sum_{i=1,2} \nabla Y(\nabla u_i, \nabla u_i) = 0.$$

We compute the integrand involving u_1 and u_2 respectively. Note the following formulas:

$$\nabla Y(\nabla u_1, \nabla u_1) = \sum_{i,j=1,\dots,n} \left[\frac{\partial}{\partial x_j}\left(Y_i \frac{\partial u_1}{\partial x_i} \frac{\partial u_1}{\partial x_j}\right) - Y_i \frac{\partial^2 u_1}{\partial x_i \partial x_j} \frac{\partial u_1}{\partial x_j} - Y_i \frac{\partial^2 u_1}{\partial x_j \partial x_j} \frac{\partial u_1}{\partial x_i}\right].$$

After an integration by parts, the first term in the right-hand side can be transformed into an integration on the free boundary.

We also have

$$\nabla\left(\frac{1}{2}|\nabla u_1|^2 - F_1(u_1)\right) \cdot Y = \sum_{i,j=1,\dots,n} Y_i \frac{\partial^2 u_1}{\partial x_i \partial x_j} \frac{\partial u_1}{\partial x_j} - \sum_{i=1,\dots,n} f_1(u_1) Y_i \frac{\partial u_1}{\partial x_i}.$$

Now, by integration by parts we get

$$\int_{B_1(0)\cap\{u_1>0\}} \left(\frac{1}{2}|\nabla u_1|^2 - F_1(u_1)\right) \operatorname{div} Y - \nabla Y(\nabla u_1, \nabla u_1)$$

$$= \int_{B_1(0)\cap\partial\{u_1>0\}} \left(\frac{1}{2}|\nabla u_1|^2 - F_1(u_1)\right) Y \cdot \nu$$

$$- \int_{B_1(0)\cap\{u_1>0\}} \nabla\left(\frac{1}{2}|\nabla u_1|^2 - F_1(u_1)\right) Y$$

$$- \int_{B_1(0)\cap\partial\{u_1>0\}} \sum_{i,j=1,\dots,n} Y_i \frac{\partial u_1}{\partial x_i} \frac{\partial u_1}{\partial x_j} \nu_j$$

$$+ \int_{B_1(0)\cap\{u_1>0\}} \sum_{i,j=1,\dots,n} Y_i \frac{\partial^2 u_1}{\partial x_i \partial x_j} \frac{\partial u_1}{\partial x_j} + \sum_{i=1,\dots,n} Y_i \frac{\partial u_1}{\partial x_i} \Delta u_1$$

$$= -\frac{1}{2} \int_{B_1(0)\cap\partial\{u_1>0\}} |\nabla u_1|^2 Y \cdot \nu.$$

Here v is the outward unit normal vector field to $\partial\{u_1 > 0\}$. In the last equality, we use the fact in $\{u_1 > 0\}$, $-\Delta u_1 = f_1(u_1)$, and in the integration on the free boundary, we have

$$\nabla u_1 = -|\nabla u_1|v, \quad \text{on } \partial\{u_1 > 0\},$$

and

$$F_1(u_1) = F_1(0) = 0 \quad \text{on } \partial\{u_1 > 0\}.$$

The calculation of u_2 is similar, and the result only differs by a sign, due to our choice of v:

$$\int_{B_1(0)\cap\{u_2>0\}} \left(\frac{1}{2}|\nabla u_2|^2 - F_2(u_2)\right) \operatorname{div} Y - \nabla Y(\nabla u_2, \nabla u_2)$$

$$= \frac{1}{2}\int_{B_1(0)\cap\partial\{u_2>0\}} |\nabla u_2|^2 Y \cdot v.$$

Adding these two, we get

$$\int_{B_1(0)\cap\partial\{u_1>0\}\cap\partial\{u_2>0\}} (|\nabla u_1|^2 - |\nabla u_2|^2)Y \cdot v = 0.$$

In view of the arbitrary choice of Y, we then get

$$|\nabla u_1| = |\nabla u_2|, \quad \text{on } B_1(0) \cap \partial\{u_1 > 0\} \cap \partial\{u_2 > 0\}. \qquad \square$$

Remark 7.1.5 In this proposition, the stronger regularity assumption is unnecessary– we only need the Lipschitz regularity of the regular part of the free boundary. Because even only with this lower regularity, we can still integrate by parts on the domain. Here the normal derivative $\frac{\partial u_i}{\partial v}$ on the free boundary must be taken in the almost everywhere sense (with respect to the $n - 1$ dimensional Hausdorff measure on the free boundary), and the claim in the proposition should reads

$$|\nabla u_i| = |\nabla u_j|, \quad H^{n-1} \quad \text{a.e. on } \partial\{u_i > 0\} \cap \partial\{u_j > 0\}.$$

This can still ensure that the Radon measure parts of Δu_1 and Δu_2 coincide.

7.2 The Parabolic Case

In this section we consider the parabolic analogue of (7.1):

$$\frac{\partial u_i}{\partial t} - \Delta u_i = f_i(u_i) - \kappa u_i \sum_{j\neq i} b_{ij}u_j^2, \quad \text{in } B_1(0) \times (-1, 1). \qquad (7.9)$$

Inspired by the elliptic case, the singular limit as $\kappa \to +\infty$ should be

$$\begin{cases} \dfrac{\partial u_i}{\partial t} - \Delta u_i \leq f_i(u_i), & \text{in } B_1(0) \times (-1, 1), \\[2mm] \left(\dfrac{\partial u_i}{\partial t} - \Delta\right)\left(u_i - \displaystyle\sum_{j \neq i} u_j\right) \geq f_i(u_i) - \displaystyle\sum_{j \neq i} f_j(u_j), & \text{in } B_1(0) \times (-1, 1), \\[2mm] u_i \geq 0, & \text{in } B_1(0) \times (-1, 1), \\[2mm] u_i u_j = 0, & \text{in } B_1(0) \times (-1, 1). \end{cases}$$
$$(7.10)$$

If we consider the initial-boundary value problems of (7.9), we can derive a standard energy inequality, since this parabolic equation is the decreasing gradient flow of the functional (7.2). This energy inequality passes to the limit $\kappa \to +\infty$ (see [5]). So, without loss of generality, we can assume

$$\int\int_{Q_1(0)} |\nabla u_\kappa|^2 + \left|\frac{\partial u_\kappa}{\partial t}\right|^2 \leq C, \qquad (7.11)$$

and for any $t \in (-1, 1)$

$$\int_{B_1 \times \{t\}} |\nabla u_\kappa|^2 + \kappa \sum_{i \neq j} u_{i,\kappa}^2 u_{j,\kappa}^2 \leq C, \qquad (7.12)$$

for some constant $C > 0$ independent of κ. The result reads as the following theorem.

Theorem 7.2.1 *As $\kappa \to +\infty$, any bounded solution u_κ of (7.9) converges to u, which satisfies the system* (7.10).

Our proof needs the assumption that the regular part of the free boundary are sufficiently smooth ($C^{2+\alpha, 1+\frac{\alpha}{2}}$ is sufficient). This has been proved (after the completion of this thesis) by Tavares and Terracini [42] (in the elliptic case) and by the author [21] (in the parabolic case) in collaboration with Dancer and Zhang. There a partial regularity theory for the free boundary problem in the singular limit was also developed. We refer the reader to these papers for more details.

The proof is similar to the elliptic case in spirit, that is, we try to derive the limit equation from a local monotonicity formula. But because in parabolic case, we do not know whether there is the analogue of (7.4), we will use another approach.

For any fixed $\kappa < +\infty$, consider a bounded solution u_κ of (7.9). It is smooth, too.

Since we assume a uniform bound on u_κ, similar to the method of [36], we can prove the uniform Hölder continuity of u_κ with respect to the parabolic distance.[2] Thus, without loss of generality, we can assume that u_κ converges to u uniformly on $Q_1(0)$.

---

[2]For a proof, see [20].

From (7.10) and (7.11), we can assume

$$\nabla u_\kappa \rightharpoonup \nabla u, \quad \text{in } L^2(Q_1(0)),$$

$$\frac{\partial u_\kappa}{\partial t} \rightharpoonup \frac{\partial u}{\partial t}, \quad \text{in } L^2(Q_1(0)).$$

In fact the first one can be improved to be a strong convergence [5].
Moreover, we have (cf. [5])

$$\lim_{\kappa \to +\infty} \int \int_{Q_1} \kappa \sum_{i \neq j} u_{i,\kappa}^2 u_{j,\kappa}^2 = 0.$$

The limit function u satisfies

$$\begin{cases} \dfrac{\partial u_i}{\partial t} - \Delta u_i = f_i(u_i), & \text{in } \{u_i > 0\}, \\ u_i \geq 0, & \text{in } B_1(0) \times (-1,1), \\ u_i u_j = 0, & \text{in } B_1(0) \times (-1,1). \end{cases}$$

We will derive a standard monotonicity formula, which was already pointed out in [5]. For any fixed $(x_0, t_0) \in Q_1(0)$, define the backward heat kernel:

$$G(x,t) = \left(4\pi |t - t_0|\right)^{-\frac{n}{2}} e^{-\frac{|x-x_0|^2}{4|t-t_0|}}.$$

In the calculation of monotonicity formula we will simply take (x_0, t_0) to be the origin $(0,0)$.

Take a $\varphi \in C_0^\infty(B_1(0))$. Define

$$D_\kappa(r) = \int_{-4r}^{-r} \int_{\mathbb{R}^n} \left[\frac{1}{2} \sum_i |\nabla u_{i,\kappa}|^2 + \frac{\kappa}{4} \sum_{i \neq j} u_{i,\kappa}^2 u_{j,\kappa}^2 \right] \varphi^2 G(x,t) \, dx \, dt.$$

Concerning u, we have a similar definition:

$$D(r) = \int_{-4r}^{-r} \int_{\mathbb{R}^n} \frac{1}{2} \sum_i |\nabla u_i|^2 \varphi^2 G(x,t) \, dx \, dt.$$

The calculation of the derivative of $D_\kappa(r)$ is standard. However, since below we need to calculate the derivative of $D(r)$, we will repeat the calculation here for clarity.

Firstly, by changing the coordinates through

$$x = r^{\frac{1}{2}} y, \qquad t = rs,$$

we get

$$D_\kappa(r) = r \int_{-4}^{-1} \int_{\mathbb{R}^n} \left[\frac{1}{2} |\nabla u(r^{\frac{1}{2}} y, rs)|^2 + H(u(r^{\frac{1}{2}} y, rs)) \right] \varphi(r^{\frac{1}{2}} y)^2 G(y,s) \, dy \, ds.$$

Here we abbreviate the index i and κ, and $H(u) = \frac{\kappa}{4} \sum_{i \neq j} u_{i,\kappa}^2 u_{j,\kappa}^2$.
Define

$$u^r(y, s) = u(r^{\frac{1}{2}} y, rs),$$

then the above formula can be written as

$$D_\kappa(r) = \int_{-4}^{-1} \int_{\mathbb{R}^n} \left[\frac{1}{2} |\nabla u^r(y, s)|^2 + rH(u^r(y, s))^2 \right] \varphi(r^{\frac{1}{2}} y)^2 G(y, s) \, dy \, ds.$$

So

$$D_\kappa'(r) = \int_{-4}^{-1} \int_{\mathbb{R}^n} \left[\nabla u^r \cdot \nabla \frac{\partial u^r}{\partial r} + r \frac{\partial H}{\partial u} \frac{\partial u^r}{\partial r} + H(u^r) \right] \varphi(r^{\frac{1}{2}} y)^2 G(y, s)$$

$$+ \left[\frac{1}{2} |\nabla u^r(y, s)|^2 + rH(u^r(y, s))^2 \right] r^{-\frac{1}{2}} \varphi(r^{\frac{1}{2}} y) \nabla \varphi(r^{\frac{1}{2}} y) \cdot y G(y, s).$$

Through an integration by parts, the term involving $\nabla u^r(y, s) \cdot \nabla \frac{\partial u^r}{\partial r}$ can be transformed into

$$- \int_{-4}^{-1} \int_{\mathbb{R}^n} \Delta u^r \frac{\partial u^r}{\partial r} \varphi(r^{\frac{1}{2}} y)^2 G(y, s) + \frac{\partial u^r}{\partial r} \nabla u^r \cdot \nabla G(y, s) \varphi(r^{\frac{1}{2}} y)^2$$

$$+ \frac{\partial u^r}{\partial r} \nabla u^r \cdot \nabla \left(\varphi(r^{\frac{1}{2}} y)^2 \right) G(y, s).$$

Note that

$$\nabla G(y, s) = \frac{y}{2s} G(y, s),$$

$$\frac{\partial u^r}{\partial r}(y, s) = \frac{1}{2} r^{-\frac{1}{2}} \nabla u \cdot y + s \frac{\partial u}{\partial t} (r^{\frac{1}{2}} y, rs),$$

$$\Delta u^r(y, s) = r \Delta u(r^{\frac{1}{2}} y, rs),$$

and the equation for u is (now write the index κ explicitly)

$$\partial_t u_{i,\kappa} - \Delta u_{i,\kappa} + \frac{\partial H}{\partial u_{i,\kappa}} = f_i(u_{i,\kappa}).$$

Substituting these into the above formula and coming back to (x, t) coordinates, we get

$$D_\kappa'(r) = - \int_{-4r}^{-r} \int_{\mathbb{R}^n} \frac{1}{2t} \sum_i \left| \nabla u_{i,\kappa} \cdot x + 2t \frac{\partial u_{i,\kappa}}{\partial t} \right|^2 \varphi^2 G$$

$$+ \int_{-4r}^{-r} \int_{\mathbb{R}^n} \sum_i f_i(u_{i,\kappa}) \left(\nabla u_{i,\kappa} \cdot x + 2t \frac{\partial u_{i,\kappa}}{\partial t} \right) \varphi^2 G$$

$$+ \int_{-4r}^{-r} \int_{\mathbb{R}^n} \frac{\kappa}{2} \sum_{i \neq j} u_{i,\kappa}^2 u_{j,\kappa}^2 \varphi^2 G$$

$$- 2 \int_{-4r}^{-r} \int_{\mathbb{R}^n} \sum_i \left(\nabla u_{i,\kappa} \cdot x + 2t \frac{\partial u_{i,\kappa}}{\partial t} \right) \nabla u_{i,\kappa} \cdot \nabla \varphi \varphi G$$

$$+ \int_{-4r}^{-r} \int_{\mathbb{R}^n} \left[\sum_i |\nabla u_{i,\kappa}|^2 + \frac{\kappa}{2} \sum_{i \neq j} u_{i,\kappa}^2 u_{j,\kappa}^2 \right] \nabla \varphi \cdot x \varphi G.$$

By letting $\kappa \to +\infty$, and noting the weak convergence of ∇u_κ and $\frac{\partial u_\kappa}{\partial t}$, we get

$$D'(r) \leq - \int_{-4r}^{-r} \int_{\mathbb{R}^n} \frac{1}{2t} \sum_i \left| \nabla u_i \cdot x + 2t \frac{\partial u_i}{\partial t} \right|^2 \varphi^2 G$$

$$+ \int_{-4r}^{-r} \int_{\mathbb{R}^n} \sum_i f_i(u_i) \left(\nabla u_i \cdot x + 2t \frac{\partial u_i}{\partial t} \right) \varphi^2 G$$

$$- 2 \int_{-4r}^{-r} \int_{\mathbb{R}^n} \sum_i \left(\nabla u_i \cdot x + 2t \frac{\partial u_i}{\partial t} \right) \nabla u_i \cdot \nabla \varphi \varphi G$$

$$+ \int_{-4r}^{-r} \int_{\mathbb{R}^n} \left[\sum_i |\nabla u_i|^2 \right] \nabla \varphi \cdot x \varphi G. \qquad (7.13)$$

Note that, the inequality only arises from the first term of the right-hand side.

With this monotonicity formula for u, we can establish the analogue theory of [4] to prove the (partial) regularity of u and its free boundaries. We will assume the regular part of the free boundary is sufficiently smooth, for example, $C^{2+\alpha,1+\frac{\alpha}{2}}$. Then, $\frac{\partial u}{\partial t}$ can be continuously extended to the regular part of the free boundary.

In order to prove our main result, as in the elliptic case, we only need to prove:

Proposition 7.2.2 *On the regular part of* $\partial \{u_i > 0\} \cap \partial \{u_j > 0\}$

$$|\nabla u_i| = |\nabla u_j|.$$

Remark 7.2.3 As before, since the regular part of the free boundary is $C^{2+\alpha,1+\frac{\alpha}{2}}$, u_i are $C^{1,\alpha}$ continuous up to the regular part of the free boundaries, and it makes sense to say the value of ∇u_i on $\partial \{u_i > 0\}$.

Proof Suppose in $Q_1(0)$, there are only two components u_1 and u_2 non-vanishing, and the free boundary $\partial \{u_i > 0\} \cap \partial \{u_j > 0\}$ is a sufficiently smooth hypersurface. This can be achieved by rescaling a sufficiently small cylinder. Moreover, we can assume the free boundary is almost flat.

Let us repeat the calculation of $D'(r)$ after the above calculation of $D'_\kappa(r)$. The first difficulty appears in differentiating

$$D(r) = \int_{-4}^{-1} \int_{\mathbb{R}^n} \frac{1}{2}\big(|\nabla u_1^r(y,s)|^2 + |\nabla u_2^r(y,s)|^2\big)\varphi(r^{\frac{1}{2}}y)^2 G(y,s)\,dy\,ds,$$

because ∇u_i^r ($i = 1, 2$) are not differentiable functions on the entire space (there is a jump when crossing the free boundary). So we need to take another view, that is, to write the above formula as

$$D(r) = \int_{-4}^{-1} \int_{\mathbb{R}^n \cap \{u_1^r > 0\}} \frac{1}{2}|\nabla u_1^r(y,s)|^2 \varphi(r^{\frac{1}{2}}y)^2 G(y,s)\,dy\,ds$$

$$+ \int_{-4}^{-1} \int_{\mathbb{R}^n \cap \{u_2^r > 0\}} \frac{1}{2}|\nabla u_2^r(y,s)|^2 \varphi(r^{\frac{1}{2}}y)^2 G(y,s)\,dy\,ds.$$

However, here the domain of integration also varies as r changes. When differentiating in r, the variation of the domain will introduce a term with the form:

$$\int_{-4r}^{-r} \int_{\mathbb{R}^n \cap \partial\{u_1>0\} \cap \partial\{u_2>0\}} (|\nabla u_1|^2 - |\nabla u_2|^2)\varphi^2 G\Theta\,d\sigma. \qquad (7.14)$$

Here $d\sigma$ is the area measure on $\partial\{u_1 > 0\} \cap \partial\{u_2 > 0\}$(note that it is a regular hypersurface), and Θ is a negative function on $\partial\{u_1 > 0\} \cap \partial\{u_2 > 0\}$ which only depends the geometry of the free boundary. The sign before $|\nabla u_1|^2$ and $|\nabla u_2|^2$ are related to the fact that the origin lies in $\{u_1 > 0\}$, otherwise the sign will be reversed.

The differentiation of the integrand can be proceeded as usual. The calculation is similar to the smooth case. There is only a new term appearing when we integrate the following by parts:

$$\int \int \nabla u^r \nabla \frac{\partial u^r}{\partial r}\varphi^2 G.$$

A boundary term will appear after integration by parts (other terms are integrations in the domain $\{u_1 > 0\}$ and $\{u_2 > 0\}$, which are exactly the same to those in the calculation of $D'_\kappa(r)$):

$$\int_{-4r}^{-r} \int_{\mathbb{R}^n \cap \partial\{u_1>0\} \cap \partial\{u_2>0\}} \left[\frac{\partial u_1}{\partial \nu}\left(\nabla u_1(x) \cdot \frac{x}{|x|} + 2t\frac{\partial u_1}{\partial t}\right) \right.$$

$$\left. - \frac{\partial u_2}{\partial \nu}\left(\nabla u_2(x) \cdot \frac{x}{|x|} + 2t\frac{\partial u_2}{\partial t}\right) \right]\varphi^2 G. \qquad (7.15)$$

Here ν is the inward unit normal vector to $\{u_1 > 0\}$ in the space direction.

Note that on $\partial\{u_1 > 0\} \cap \partial\{u_2 > 0\}$

$$\nabla u_1 = |\nabla u_1|\nu, \qquad \nabla u_2 = -|\nabla u_2|\nu.$$

So (7.15) is

$$\int_{-4r}^{-r} \int_{\mathbb{R}^n \cap \partial \{u_1 > 0\} \cap \partial \{u_2 > 0\}} \left[\left(|\nabla u_1|^2 - |\nabla u_2|^2 \right) v \cdot \frac{x}{|x|} \right.$$

$$\left. + 2t \left(|\nabla u_1| \frac{\partial u_1}{\partial t} - |\nabla u_2| \frac{\partial u_2}{\partial t} \right) \right] \varphi^2 G. \qquad (7.16)$$

Now substituting (7.14) and (7.16) into (7.13), we get (inserting the dependence on (x_0, t_0), which was omitted before)

$$0 \geq \int_{-4r}^{-r} \int_{\mathbb{R}^n \cap \partial \{u_1 > 0\} \cap \partial \{u_2 > 0\}} \left(|\nabla u_1|^2 - |\nabla u_2|^2 \right) \Theta(x - x_0, t - t_0) \varphi^2 G(x - x_0, t - t_0)$$

$$+ \left(|\nabla u_1|^2 - |\nabla u_2|^2 \right) v \cdot \frac{x - x_0}{|x - x_0|} \varphi^2 G(x - x_0, t - t_0)$$

$$+ 2(t - t_0) \left(|\nabla u_1| \frac{\partial u_1}{\partial t} - |\nabla u_2| \frac{\partial u_2}{\partial t} \right) \varphi^2 G(x - x_0, t - t_0).$$

We are free to choose φ and (x_0, t_0). r can also be arbitrary small. From this, arguing by contradiction (note that the integrand are all continuous functions on the free boundary), we see

$$0 \geq \left(|\nabla u_1|^2 - |\nabla u_2|^2 \right) \Theta(x - x_0, t - t_0) + \left(|\nabla u_1|^2 - |\nabla u_2|^2 \right) v \cdot \frac{x - x_0}{|x - x_0|}$$

$$+ 2(t - t_0) \left(|\nabla u_1| \frac{\partial u_1}{\partial t} - |\nabla u_2| \frac{\partial u_2}{\partial t} \right).$$

Now it is easy to derive our conclusion. First, we can choose $t = t_0$, while (x_0, t_0) is not on the free boundary, so the last term vanishes. Then noticing the fact that $(x_0, t_0) \in \{u_1 > 0\}$, so

$$v \cdot \frac{x - x_0}{|x - x_0|} < 0.$$

This choice also ensures $\Theta < 0$. Then the above inequality implies, on the free boundary

$$|\nabla u_1| \geq |\nabla u_2|.$$

Since it should be symmetric with respect to u_1 and u_2, the above inequality must be an equality.   □

References

1. Caffarelli, L.A., Cabre, X.: Fully Nonlinear Elliptic Equations. American Mathematical Society Colloquium Publications, vol. 43. Am. Math. Soc., Providence (1995)
2. Caffarelli, L.A., Lin, F.: An optimal partition problem for eigenvalues. J. Sci. Comput. **31**(1), 5–18 (2007)
3. Caffarelli, L.A., Lin, F.: Singularly perturbed elliptic systems and multi-valued harmonic functions with free boundaries. J. Am. Math. Soc. **21**(3), 847–862 (2008)
4. Caffarelli, L.A., Lin, F.: Nonlocal heat flows preserving the L^2 energy. Discrete Contin. Dyn. Syst., Ser. A **23**(1–2), 49–64 (2009)
5. Caffarelli, L.A., Karakhanyan, A.L., Lin, F.: The geometry of solutions to a segregation problem for non-divergence systems. J. Fixed Point Theory Appl. **5**(2), 319–351 (2009)
6. Chen, X.-Y.: A strong unique continuation theorem for parabolic equations. Math. Ann. **311**(4), 603–630 (1998)
7. Conti, M., Terracini, S., Verzini, G.: An optimal partition problem related to nonlinear eigenvalues. J. Funct. Anal. **198**(1), 160–196 (2003)
8. Conti, M., Terracini, S., Verzini, G.: A variational problem for the spatial segregation of reaction diffusion systems. Indiana Univ. Math. J. **54**(3), 779–815 (2005)
9. Conti, M., Terracini, S., Verzini, G.: Asymptotic estimates for the spatial segregation of competitive systems. Adv. Math. **195**(2), 524–560 (2005)
10. Conti, M., Terracini, S., Verzini, G.: Uniqueness and least energy property for strongly competing systems. Interfaces Free Bound. **8**, 437–446 (2006)
11. Dancer, E.N., Du, Y.H.: Positive solutions for a three-species competition system with diffusion-I. general existence results. Nonlinear Anal. **24**(3), 337–357 (1995)
12. Dancer, E.N., Du, Y.H.: Positive solutions for a three-species competition system with diffusion—II: the case of equal birth rates. Nonlinear Anal. **24**(3), 359–373 (1995)
13. Dancer, E.N., Zhang, Z.: Dynamics of Lotka–Volterra competition systems with large interaction. J. Differ. Equ. **182**(2), 470–489 (2002)
14. Dancer, E.N., Wang, K., Zhang, Z.: Uniform Hölder estimate for singularly perturbed parabolic systems of Bose–Einstein condensates and competing species. J. Differ. Equ. **251**, 2737–2769 (2011)
15. Dancer, E.N., Wang, K., Zhang, Z.: Dynamics of strongly competing systems with many species. Trans. Am. Math. Soc. **364**, 961–1005 (2012)
16. Dancer, E.N., Wang, K., Zhang, Z.: The limit equation for the Gross–Pitaevskii equations and S. Terracini's conjecture. J. Funct. Anal. **262**(2), 1087–1131 (2012)
17. Gilbarg, D., Trudinger, N.S.: Elliptic Partial Differential Equations of Second Order. Springer, Berlin (2001)
18. Gromov, M., Schoen, R.: Harmonic maps into singular spaces and p-adic superrigidity for lattices in groups of rank one. Publ. Math. IHÉS **76**(1), 165–246 (1992)

19. Hall, D.S., Matthews, M.R., Ensher, J.R., Wieman, C.E., Corne, E.A.: Dynamics of component separation in a binary mixture of Bose–Einstein condensates. Phys. Rev. Lett. **81**(8), 1539–1542 (1998)
20. Han, Q., Lin, F.: Nodal sets of solutions of elliptic differential equations. Books available on Han's homepage
21. Ladyzenskaja, O.A., Solonnikov, V.A., Ural'ceva, N.N.: Linear and Quasilinear Equations of Parabolic Type. Translations of Mathematical Monographs, vol. 23. Am. Math. Soc., Providence (1968)
22. Leung, A.W.: Systems of Nonlinear Partial Differential Equations: Applications to Biology and Engineering. Kluwer Academic, Dordrecht (1989)
23. Lieberman, G.M.: Second Order Parabolic Differential Equations. World Scientific, Singapore (2005)
24. Lin, F., Yang, X.: Geometric Measure Theory: An Introduction. Science Press/International Press, Beijing/Boston (2002)
25. Noris, B., Tavares, H., Terracini, S., Verzini, G.: Uniform Hölder bounds for nonlinear Schrödinger systems with strong competition. Commun. Pure Appl. Math. **63**(3), 267–302 (2010)
26. Noris, B., Tavares, H., Terracini, S., Verzini, G.: Convergence of minimax and continuation of critical points for singularly perturbed systems. Preprint
27. Poon, C.C.: Unique continuation for parabolic equations. Commun. Partial Differ. Equ. **21**(3–4), 521–539 (1996)
28. Smale, S.: On the differential equations of species in competition. J. Math. Biol. **3**, 5–7 (1976)
29. Smith, H.L.: Monotone Dynamical Systems: An Introduction to the Theory of Competitive and Cooperative Systems. Mathematical Surveys and Monographs. Am. Math. Soc., Providence (1995)
30. Tavares, H., Terracini, S.: Regularity of the nodal set of segregated critical configurations under a weak reflection law. Calc. Var. (2011). doi:10.1007/s00526-011-0458-z
31. Timmermans, E.: Phase separation of Bose–Einstein condensates. Phys. Rev. Lett. **81**(26), 5718–5721 (1998)

Index

K. Wang, *Free Boundary Problems and Asymptotic Behavior of Singularly
Perturbed Partial Differential Equations*, Springer Theses, DOI 10.1007/978-3-642-33696-6,
© Springer-Verlag Berlin Heidelberg 2013